T0230893

Fruit Flies
and the
Sterile Insect Technique

Edited by

Carrol O. Calkins

Waldemar Klassen

Pablo Liedo

CRC Press
Taylor & Francis Group
Boca Raton London New York

CRC Press is an imprint of the
Taylor & Francis Group, an **informa** business

First published 1994 by CRC Press
Taylor & Francis Group
6000 Broken Sound Parkway NW, Suite 300
Boca Raton, FL 33487-2742

Reissued 2018 by CRC Press

Library of Congress Cataloging-in-Publication Data

Fruit flies and the sterile insect technique / edited by Carrol O.
    Calkins, Waldemar Klassen, Pablo Liedo.
        p.   cm.
        Papers from the International Congress of Entomology, held in
    Beijing, China, June 27-July 4, 1992.
        Includes bibliographical references and index.
        ISBN 0-8493-4854-4
        1. Fruit-flies—Biological control—Congresses.   2. Insect
    sterilization—Congresses.   I. Calkins, Carrol O.   II. Klassen,
    Waldemar.   III. Liedo, Pablo.   IV. International Congress of
    Entomology (1992 : Peking, China)
    SB945.F8F75   1994
    632'.746—dc20   94-13258

Publisher's Note
The publisher has gone to great lengths to ensure the quality of this reprint but points out that some imperfections in the original copies may be apparent.

Disclaimer
The publisher has made every effort to trace copyright holders and welcomes correspondence from those they have been unable to contact.

ISBN 13: 978-1-315-89306-8 (hbk)
ISBN 13: 978-1-351-07216-8 (ebk)

Visit the Taylor & Francis Web site at http://www.taylorandfrancis.com and the
CRC Press Web site at http://www.crcpress.com

acceptable, more reliable and cheaper. Fourthly, and possibly of most importance is the concern for the use of insecticides to control or eradicate populations or new infestations. Experiences in Mexico, Guatemala and California programs suggest that the public may no longer tolerate extensive aerial applications of bait sprays involving chemical insecticides, especially in urban and suburban areas. Therefore, the expanded use of the SIT will create a very great need for additional information and refinements in technology.

The editors would like to acknowledge the support of the Organizing Committee of the International Congress Of Entomology in Beijing, China, the initiative and hospitality of Dr. Wang Huasong, of the International Organization for Biological Control (IOBC) through the Global Working Group on Fruit Flies, the support of the Entomological Society of America for travel funds for certain individuals, the financial and moral support of our respective agencies and finally for the financial support of the Agricultural Research Service, U. S. Department of Agriculture for the Proceedings.

The international community of fruit fly workers has a long history of collaborative and cooperative efforts. This particular Symposium represents the continuation of this productive collaboration. Free and open discussions of results and ideas are essential for scientific progress, and meetings like this provide the opportunity for such discussions. These proceedings will be an addition to a series of books that have followed other meetings of this type. These books are very useful for both the new and old specialists and for the nonspecialists looking for an overview and the state of the art of the science of fruit flies.

C. O. Calkins
W. Klassen
P. Liedo

PREFACE

The Sterile Insect Technique (SIT) was conceived by Dr. E. F. Knipling as an ecologically selective measure for the area-wide management of insect pest populations, and for eradication from areas surrounded by natural or man-made barriers sufficiently effective to prevent reinvasion except through the intervention of man. The advent of the nuclear age provided the tool (gamma radiation) to sterilize, cheaply and easily, large numbers of reared insects for release against a wild population. The development and application of the SIT has necessitated a great deal of research on the taxonomy, behavior, genetics, physiology, nutrition, population dynamics, bioclimatology, and economic aspects of target pests. The need for so much information stems from the fact that in employing the sterile insect technique, we are using a member of an insect species to strongly suppress its own conspecific wild populations in all niches on an area-wide basis. Millions of individual insects must be reared, sexually sterilized, and released into the field each week. All of the released sterile males must be distributed so that they may successfully search out the niches where they are likely to encounter wild females. They must be successful in competing with wild males for wild females, in courtship and in transferring irradiated sperm.

The Sterile Insect Technique was first used on an operational basis against the New World screwworm in the Southeastern United States in 1957-1959. Subsequently, the technique has been used successfully against the Mediterranean fruit fly, melon fly, Mexican fruit fly, pink bollworm, codling moth, tsetse flies, the onion fly and against several species of mosquitoes. For the technique to be successful, it has been necessary to allocate several million dollars each year for research and methods development on the basic biology and behavior of these species. The need to acquire new information about these species continues to be essential to the success of these control and eradication programs. With regard to the costs of operating an SIT program, 30 to 50 percent of the cost is incurred in finding out what is happening in the field than what is incurred in producing and releasing the insects.

The primary focus of this Symposium is on tephritid fruit flies. We anticipate that during the next decade the use of the sterile insect technique against tephritid fruit flies will be expanded significantly. This prediction is based on several factors. First, the leaders of the fruit and vegetable industries in many countries have become acutely aware that the profitability of these industries depends on exporting fruit and vegetables to lucrative markets in developed countries that will not tolerate the introduction of these pests. Secondly, the highly successful programs in Australia, Mexico, Japan and elsewhere have proven that the SIT can be used effectively under very difficult circumstances. Thirdly, the development of new technologies including genetic sexing stains and better attractants will make the SIT more

THE EDITORS

Carrol O. Calkins presently is Research Leader/Location Coordinator at the Fruit and Vegetable Insect Research Laboratory, U. S. Department of Agriculture (USDA) in Yakima, Washington where he supervises research on codling moth, pear psylla, Colorado potato beetle and green peach aphid. From 1972 to 1977 and from 1980 to 1993, Dr. Calkins was Research Entomologist and Research Leader at the Insect Attractants, Behavior and Basic Biology Research Laboratory, USDA, Gainesville, Florida. During the early period he studied the ecology and behavior of the plum curculio. During the latter period he was involved in fruit fly behavior, especially quality control of fruit flies mass reared for sterile release programs as well as mating behavior and biological control of Caribbean fruit flies. He coordinated much of the fruit fly research within the laboratory. From 1977 to 1980 he served as Head of the Entomology section of the Seibersdorf Laboratories, International Atomic Energy Agency, Vienna, Austria where the technology for mass rearing of Mediterranean fruit fly and tsetse flies was developed. From 1964 to 1972, Dr. Calkins was Research Entomologist studying the ecology, biology and control of corn rootworms, false wireworms and army cutworms at the Northern Grain Insects Research Laboratory, USDA, Brookings, South Dakota. He first began with the U. S. Department of Agriculture at Lincoln, Nebraska in 1960 where he worked on host plant resistance of alfalfa to the spotted alfalfa aphid and the pea aphid and the behavioral ecology of the sweetclover weevil.

Dr. Calkins serves on the Science Advisory Panel for Medfly Eradication Programs for the California Department of Food and Agriculture. He is presently on the International Affairs Committee of the Entomological Society of America. He was Chairman from 1984 to 1990. He has served as Chairman of the IOBC Global Working Group on Quality Control of Mass Reared Insects from 1984 through 1988. He has served on the Caribbean Fruit Fly Protocol Committee for the Development of Fly-free Zones and on the Caribbean Fruit Fly Technical Committee for the Florida Department of Agriculture.

Waldemar Klassen is Deputy Director, Joint FAO/IAEA Division of Nuclear Techniques in Food and Agriculture, Vienna, Austria. From 1990 to 1992 Dr. Klassen served in this Division as Head, Insect and Pest Control Section. In this capacity he was responsible for Coordinated Research Programmes and Technical Cooperation Projects on control and eradication of tropical fruit flies, tsetse flies, the New World screwworm and various harmful caterpillars. From 1988 to 1990 he served as Associate Deputy Administrator for Plant Sciences and Natural Resources, Agricultural Research Service, U. S. Department of Agriculture. From 1983 to 1988 Dr. Klassen served as Director of the Beltsville Area of this Service. In this capacity he launched a program to upgrade the facilities and scientific programs of the Beltsville Agricultural Research Center and U. S. National Arboretum. From 1980 to 1983 he served as National Program Director, Crop Protection Sciences, and from 1972 to 1980 as the National Program Leader for Pest Management, National Program Staff, Agricultural Research Service. From 1965 to 1967 he served as Research Geneticist (Insects) and until 1971 as Research Leader, Insect Physiology and Metabolism, Metabolism and Radiation Research Laboratory, Fargo, North Dakota. From 1963 to 1965 he conducted postdoctoral research in insect genetics, University of Illinois, Urbana. He was awarded the Ph.D. in 1963 by the University of Western Ontario, London, Ontario, Canada.

Dr. Klassen serves on the Council of the International Congresses of Entomology and the Standing Committee for the International Plant Protection Congresses. He received the Meritorious Rank Executive Award from the President of the United States, and he is a Fellow of the Entomological Society of America.

Pablo Liedo is Researcher of Insect Ecology and Pest Management in the "Centro de Investigaciones Ecologicas del Sureste" (CIES) located in Tapachula, Chiapas, Mexico. Dr. Liedo received his B. S. in Agronomy in 1978 from the Monterrey Institute of Technology in Mexico; his M. S. in Pest Management in 1983 from the University of Southampton, England; and his Ph.D. in Entomology from the University of California at Davis in 1989. From 1979 to 1989, he joined the campaign against the Mediterranean fruit fly in Mexico and Guatemala where the world's largest mass rearing and sterilization facility exists to prevent the introduction of the Medfly from Guatemala into Mexican territory by means of the Sterile Insect Technique. In 1990 he was appointed Researcher and Area Director of the Tapachula Unit at CIES.

Dr. Liedo is a member of the Mexican and the American Entomological Societies. In 1990 he was distinguished as National Researcher in the Mexican Researchers System. Since 1989 he has been chairman of the Global Working Group on Fruit Flies of Economic Importance within the International Organization of Biological Control (IOBC).

CONTRIBUTORS

L. Baruffi
Department of Animal Biology
University of Pavia
Pavia, Italy

K. Bloem
Formerly: Research Associate
USDA-APHIS Methods
  Development
Guatemala City, Guatemala
Currently: Program Coordinator
Sterile Insect Release Program
Osoyoos, British Columbia
Canada

S. Bloem
Formerly: Research Associate
USDA-APHIS Methods
  Development
Guatemala City, Guatemala
Currently: NSERC Visiting
  Postdoctoral Fellow
  Agriculture
Canada Research Station
Summerland, British Columbia
Canada

E. J. Buyckx
Consultant in Economic
  Entomology
Joint FAO/IAEA Division of
  Nuclear Techniques in Food
  and Agriculture
International Atomic Energy
  Agency
Vienna, Austria

C. O. Calkins
Formerly: Research Leader
USDA-ARS
Insect Attractants, Behavior and
  Basic Biology Research Lab
Gainesville, Florida
Currently: Research Leader/
  Location Coordinator
USDA-ARS
Fruit and Vegetable Insect Lab
Yakima, Washington

D. Scot Campbell
Director, Operational Support
  Staff
International Services
USDA-APHIS
Hyattsville, Maryland

James R. Carey
Professor
Department of Entomology
University of California
Davis, California

D. L. Chambers
Station Director
USDA-APHIS Methods
  Development
Guatemala City, Guatemala

A. P. Economopoulos
Department Head
University of Crete
Department of Biology
Crete, Greece

Walther R. Enkerlin
Program Director
Programa Moscamed (SARH)
Tapachula, Chiapas, Mexico

Kingsley Fisher
Western Australian Department
    of Agriculture
South Perth, Western Australia

Gerald Franz
Entomologist
International Atomic Energy
    Agency
Joint FAO/IAEA Division
Entomology Unit
Seibersdorf, Austria

G. Gasperi
Geneticist
Department of Animal Biology
University of Pavia
Pavia, Italy

C. R. Guglielmino
Geneticist
Department of Genetics and
    Microbiology
University of Pavia
C.N.R. Institute of Genetics and
    Evolutionary Biology
Pavia, Italy

G. E. Haniotakis
Entomologist
NRC "Demokritos"
Department of Biology
Athens, Greece

J. Hendrichs
Formerly: Research
    Entomologist
Department of Entomology
University of Massachusetts
Amherst, Massachusetts
Currently: Entomology Unit
IAEA Laboratories
Seibersdorf, Austria

T. J. Henneberry
Laboratory Director
USDA-ARS-Western Cotton
    Research Laboratory
Phoenix, Arizona

Hiroyuki Kakinohana
Director
Okinawa Prefectural Fruit Fly
    Eradication Project
Naha, Okinawa, Japan

Philippe Kerremans
Entomologist
International Atomic Energy
    Agency
Joint FAO/IAEA Program
Entomology Unit
Seibersdorf, Austria

W. Klassen
Associate Division Director
Joint FAO/IAEA Division of
    Nuclear Techniques in Food
    and Agriculture
International Atomic Energy
    Agency
Vienna, Austria

Pablo Liedo
Research Entomologist
Centro de Investigaciones
    Ecologicas del Sureste (CIES)
Tapachula, Chiapas, Mexico

D. A. Lindquist
Head
Insect Pest Control Section
International Atomic Energy
    Agency
Vienna, Austria

A. R. Malacrida
Geneticist
Department of Animal Biology
University of Pavia
Pavia, Italy

Aldo Malavasi
Associate Professor
Department of Biology
University of Sao Paulo
Sao Paulo, SP, Brazil

A. G. Manoukas
Entomologist
NRC Demokriyos Aghia
Pazaskeri, Attiki
Athens, Greece

R. Milani
Professor
Department of Animal Biology
University of Pavia
Pavia, Italy

Dina Orozco
Factory Director
Programa Moscamed (SARH
USDA)
Tapachula, Chiapas, Mexico

R. J. Prokopy
Professor
Department of Entomology
University of Massachusetts
Amherst, Massachusetts

Jesus Reyes
Director, SAHR
Mexico City, Mexico

G. Greg Rohwer
Agricultural Consultant
Annapolis, Maryland

Y. Rossler
Research Geneticist
"Israel Cohen" Institute for
Biological Control
Rehovot, Israel

Roger I. Vargas
Research Entomologist
USDA-ARS, Tropical Fruit and
Vegetable Research Laboratory
Honolulu, Hawaii

S. Vijaysegaran
Fruit Research Division
Malaysian Agricultural Research
and Development Institute
Kuala Lumpur, Malaysia

Hua-song Wang
Entomologist
Institute for Application of
Atomic Energy
Chinese Academy of
Agricultural Sciences
Beijing, China

M. C. Zapater
Geneticist
Cat. de Genetica, Fac.
Agronomia
Univ. de Buenos Aires
Buenos Aires, Argentina

He-qin Zhang
Entomologist
Institute for Application of
Atomic Energy
Chinese Academy of
Agricultural Sciences
Beijing, China

TABLE OF CONTENTS

Fruit Flies
and the
Sterile Insect Technique

Chapter 1

OVERVIEW OF THE JOINT FAO/IAEA DIVISION'S INVOLVEMENT IN FRUIT FLY STERILE INSECT TECHNIQUE PROGRAMS

W. Klassen, D. A. Lindquist, and E. J. Buyckx

I. INTRODUCTION

A number of fruit flies in the Family Tephritidae are extremely destructive pests of fruit throughout the world. These flies have special adaptations that enable them to survive adverse conditions. The same traits that impart to them the unusual capacity to survive, also allow them to be formidable pests. In most of the important pest species, these traits include: (1) a multivoltine life cycle with explosive reproductive capacity, (2) polyphagy, or a capacity to exploit a very large number of host plants, (3) pronounced ability to disperse, and (4) ability of adults to survive up to several months of inclement weather. Not only do fruit flies cause great losses in fruit and vegetable production, but they seriously impede international trade in host commodities. Thus, they are a major impediment to economic development. Consequently, several countries had already undertaken research programs to develop the use of the sterile insect technique against fruit flies before the Joint FAO/IAEA Division for Application of Nuclear Techniques in Food and Agriculture had come into existence.

II. HISTORY

The Joint FAO/IAEA Division was formed in October 1964 from units conducting relevant research and development (R&D) in the Food and Agriculture Organization (FAO) and in the International Atomic Energy Agency (IAEA). The IAEA had assembled a panel of experts in 1962 to review available information on the sterile insect technique (SIT) and to advise the Agency on how it might encourage and support research in this field.[1] The panel recommended that the Agency should "take active steps to disseminate technical information on the application of this technique for the control or eradication of suitable insect pests such as the Mediterranean fruit fly, oriental fruit fly, olive fly, and Queensland fruit fly, as well as tsetse fly, leopard moth, tropical ox warble, cotton boll weevil, etc."[1] This was followed by a symposium in Athens, Greece, April 22-26, 1963 jointly organized by IAEA and FAO, and titled: "Radiation and Radioisotopes Applied to Insects of Agricultural Importance."[2] Moreover, a second panel was convened by the IAEA in July 1964 to review progress since 1962 and to recommend future actions.[3] This Panel recorded valuable information on

3

programs against the olive fly (*Dacus oleae* (Gmelin), Mediterranean fruit fly (medfly), *Ceratitis capitata* (Wiedemann), Queensland fruit fly, *Bactrocera tryoni* (Froggatt), oriental fruit fly, *Bactrocera dorsalis* Hendel, and the Mexican fruit fly, *Anastrepha ludens* (Loew). Thus, by the time the Joint Division was formally established, a program for the use of the sterile insect technique against fruit flies was already underway.

Several Member States had begun research on the use of the SIT against tropical fruit flies during the 1950s. Such work had been undertaken in Hawaii with the medfly, oriental fruit fly, and the melon fly, *Bactrocera cucurbitae*, and in Mexico City by the United States Department of Agriculture, Agricultural Research Service (USDA, ARS) during the mid-1950s. The main thrusts were on developing methods for population detection and monitoring, mass rearing, methods for inducing sexual sterility, and methods of distributing sterile flies.[4-7] Similar work was undertaken in Australia against the Queensland fruit fly.[8,9]

The ARS Laboratory in Hawaii had attempted to eradicate the Mediterranean fruit fly from a portion of the Island of Hawaii as early as 1959, but this proved impossible because of migration of the pest from elsewhere on the island. However, in 1963, the Hawaii Laboratory succeeded in using the SIT to eradicate the oriental fruit fly from Guam and the melon fly from the island of Rota.[10,11,7] The laboratory in Mexico City began field trials with sterile Mexican fruit flies in 1960. By 1964, the SIT was operational against the Mexican fruit fly and was used to eradicate the pest from southern California and as a quarantine measure to prevent the pest from re-entering California from Mexico.[12-14] This very successful program now protects the Rio Grande Valley of Texas from the pest.[15]

Field trials against the Queensland fruit fly began in 1962 in New South Wales.[8,16,1,3] Eradication was not achieved because of the strong migratory capacity of this species, although the population was suppressed.

In 1963 field trials with the SIT against the medfly were undertaken in Costa Rica and in Israel.[3,17] The IAEA assisted in these trials, and from 1964 to this day, the Joint FAO/IAEA Division has been involved in various ways in such programs. A chronology of field trials using the sterile insect technique against fruit fly species is given in Table 1.

III. MECHANISMS FOR ADVANCING PROGRAMS

The Joint FAO/IAEA Division has been involved in many of these programs. Seven major mechanisms are available to the Joint FAO/IAEA Division for advancing programs against these insect pests.

Table 1
Chronology of Field Trials and Operational Programs Involving
Use of the Sterile Insect Technique Against Tropical Fruit Flies

Location	Description and Results
	Mediterranean Fruit Fly, *Ceratitis capitata*
Island of Hawaii	Suppression; 31 sq. mile test area; not isolated; 1959-1960[6]
Costa Rica	Suppression in 1963 on Puntarenas peninsula by Govt. of Costa Rica, OIRSA, Inter-Amer. Inst. Agri. Sciences and Joint FAO/IAEA Division.[18]
Concordia, Argentina	Releases in 1966 in 30 ha. citrus orchard; no medflies detected for 3 years.[19]
Concordia, Argentina	Releases in 1968-1969 in 2,700 ha. citrus; excellent control.[19]
Northwest Argentina	Releases in 1973 in 60 ha. citrus orchard.[19-21]
Western Argentina	During the 1980s rearing facilities were constructed in San Juan and Mendoza Provinces. In 1991 a national program was organized to create fly-free zones.[22-24]
Capri Island, Italy	Suppression in 1967; sterile flies provided to Italian Government by Israel and Seibersdorf Lab.; technical backstopping by Joint FAO/IAEA Division.[25]
Chile	Infestations first discovered in 1964. Since 1966 infestations have been eradicated. Fly-free zone established in 1980 to permit exports to USA. Use of sterile flies began in 1983 provided by Mexico and Peru.[26-32]
Cyprus	Releases made in 1971 in small peach orchard; some suppression achieved.[33]
Procida Island, Italy	Tests in 1972-1975; 3.7 sq. km and 2.7 km from mainland; excellent suppression in commercial orchards, etc.[34-36]
Tenerife Island, Spain	Tests in 1965 through 1968; at first little effect; good control in 1968 with increased release rate.[37-38]
Alhama de Murcia, Spain	Excellent control in 1969 in citrus and stone fruit orchards. Some sterile flies from IAEA Seibersdorf Laboratory.[38,39]
Greneda, Spain	In 1972 excellent protection of peaches in isolated orchard.[40]
Caruzo, Nicaragua	In 1967 unsuccessful attempt to eradicate from 120 sq. km. Further tests in 1968-1969; 90 percent control on 48 sq. km of citrus and coffee.[41-43]
Tacna and Mosquegua, Peru	Tests in 1969-1972; variable results from year to year. Field program in 1986 and 1987. Production resumed in 1988.[44-47]
Porto-Farina, Tunisia	Tests in 1970, 1971, and 1972; 600 ha. of loquat, apricot, peach & fig; good suppression.[48]

Location	Description and Results
Southern coastal area of Israel	Tests in 1972 and 1973; good suppression; Citrus Marketing Board, UNDP, IAEA project.[49]
Lanai Island, Hawaii	Tests in 1973, 1974 & 1975; 381 sq. km; excellent suppression in spite of some invasion from Maui Island (13-16 km distant).[50,51]
Los Angeles, California	Operational program, 1975-1976; introduced infestation eradicated.[14,15]
Santa Clara Valley, California	In 1980 and 1982, operational program cost more than US $100 million; introduced infestation eradicated. Use of SIT was interrupted.[15,52,53]
Southern Mexico	Operational program began in 1979; eradicated from 15,000 sq. km.in 1982; maintained barrier to reinvasion from Guatemala from 1982 to present.[54-57]
Guatemala	Operational program began in 1984; eradication in zone bordering on Mexico and northern half of Guatemala; suppression elsewhere.[54,58]
Carnarvon, western Australia	Began in 1980; eradication achieved in 1984; subsequent reinvasions eradicated. Reinfested in 1991.[59-61]
Egypt	Between 1982 and 1986 work fundamental to eradication of the medfly from Egypt was carried out. Project was terminated before mass rearing facility could be constructed.[62]
Kauai Island, Hawaii, U.S.A.	Releases of sterile males in coffee plantations began in 1991; continuing .[63]
Southern Florida, U.S.A.	Infestation eradicated in 1985 with sterile flies shipped by air from Hawaii to Miami.[64]
Libya, valley 650 km southeast of Tripoli	Eradicated in 1991; too cold to overwinter; reintroduced each year with infested fruit.[65]
Santa Cruz de Tenerife Province, Spain	In 1975-1977 sterile flies were released on Isla de El Hierro; excellent suppression but not eradication even with 1,100 sterile flies per hectare.[66]
Israel	Excellent protection of citrus in 1989 & 1990. Sterile males from genetic sexing strain shipped from IAEA Seibersdorf Laboratory.[67]
Southern California	Eradicated in 1987.[15]
Los Angeles, California	Incipient infestation eradicated in 1981-1982 with ground application of bait sprays and sterile flies.[68]

Natal Fruit Fly,
Ceratitis rosa **(Karsch)**

Mauritius	In 1980-81, sterile flies were released, good suppression was achieved in northern part of island.[69]

Location	Description and Results

Oriental Fruit Fly,
Bactrocera dorsalis **Hendel**

Location	Description and Results
Rota Island	Tests in 1960-1962; suppression but not eradication; too few sterile flies sent from Hawaii.[6]
Guam, Saipan, Tinian, Agigugan	Tests in 1963 and 1964 with sterile flies reared in Hawaii; but eradication could be accomplished only by adding use of male annihilation.[11,70]
Ogasawara Islands, Amami Islands, Miyako Islands, Okinawa Islands, Yaeyama Islands	Operations in 1976-1986; eradication achieved on all islands; male annihilation used on all islands, and SIT was necessary on some islands.[71-75]
Taiwan	Releases in 1975-1984; effective suppression; superseded by male annihilation technique.[76]
Pat Chong, Nakorn Ratchasima Province, Pak Tho, Ratchaburi Province, Rayong, Chonburi, Province, Thailand	Program initiated in 1991.[77-78]
Guimaras Island, Philippines	Experimental releases of sterile insects, 1981-1987, and 1992.[79-80]
Ang Khang (highlands near Chang Mai), Thailand	Releases in 1984-1987; good suppression of *D. dorsalis* and *D. correctus*, but fruit damaged by uncontrolled *D. zonatus*[77,81,82]

Melon Fly,
Bactrocera cucurbitae **Coquillett**

Location	Description and Results
Rota Island	In 1962 and 1963 with sterile flies shipped from Hawaii; eradication achieved.[5,7,10,14]
Kume Island, Japan	Operations in 1975-1976; eradication achieved; however sterile flies had to be released continuously on Kume until flies eradicated from nearby neighboring islands.[71,83]
Okinawa Prefecture, Miyako Islands, Kikai Is., Amami Is., Kagoshima Prefecture, Japan	Operations began in 1979 on Amami and subsequently on all islands; by 1991 eradication achieved everywhere except Yaeyama Islands.[71,74,84-86]

Queensland Fruit Fly,
Bactrocera tryoni **(Froggatt)**

Location	Description and Results
Northeastern New South Wales, Australia	Releases made in 1961 and 1962; suppression achieved; not isolated.[8,9,87]
Perth, western Australia	Eradication achieved in 1989 and 1990 from 125 km² area.[60-61]
Cowra, western New South Wales; Victoria	Releases in 1991 and 1992; objective is to develop technology to expand suppression areas.[60-61]

Location	Description and Results
	Olive Fly, ***Dacus oleae* (Gmelin)**
Italy	Releases in 1959; very small test.[88]
Greece	Inadequate protection of fruit; inadequate quality of mass reared insects.[89]
	Chinese Citrus Fly, *Dacus citri*
Huishui County, Guizhou Province, China	Excellent suppression in 1987 and 1989 in orange groves, 34 ha.[90,91]
	Guava Fruit Fly, *Dacus correctus* (Bezzi)
Ang Khang, Chang Mai and Rayong, Chonburi Province, Thailand	Releases in 1984-1987; good suppression of *B. dorsalis* and *D. correctus*, but fruit damaged by uncontrolled *D. zonatus*.[78,79,81,82]
	Peach Fruit Fly, *Dacus zonatus*
Pakistan	Small scale releases begun in 1984 under IAEA TC Project PAK/5/018.[92,93]
	Mexican Fruit Fly, ***Anastrepha ludens* (Loew)**
Santa Rosa & San Carlos, Mexico	Releases in 1960-1962. Excellent suppression at Santa Rosa. Inadequate suppression at San Carlos.[1,94]
Tijuana and La Paz, Mexico	Releases in 1961(?). Good suppression at Tijuana; apparent eradication at La Paz.[95]
Mexico-USA border.	Since 1964, releases of sterile flies have been routinely made to prevent entry of the pest into USA. Incipient infestations have been eradicated from Arizona (1963-67), Florida (1972), and California (1983-84), and at various times from Texas.[14,15,95]
Senora, Mexico	Certified fly free zone created.[96]
Mexico	In 1991, work began on a 12 year campaign to clear northern fruit production zones of *A. ludens*. SIT will be used also for its control in central and southern Mexico.[97]
	Guava Fruit Fly, *Anastrepha striata* (Schiner), **West Indies Fruit Fly, *A. obliqua* (Macquardt),** **Sapote Fruit Fly , *A. serpentina* (Wied.)**
Mexico	In 1992, work is initiated on a 12 year campaign to clear northern fruit production zones of these species in central and southern Mexico. SIT will be used for control of the species.[97]

Location	Description and Results
	Caribbean Fruit Fly, *Anastrepha suspensa* **Loew**
Key West, Florida, USA	Releases in 1970-1972. Very strong suppression achieved.[98]
Southwest Florida	In 1988-1990 releases of sterile flies were made in a 20 km² urban area. The parasitoid, *Diachasmimorpha longicaudata*, was also released.[99]
	European Cherry Fruit Fly, *Rhagoletis cerasi* **L.**
Northwest Switzerland	Releases made in 1973 and 1974; sterile flies not fully competitive.[100,101]

A.   IAEA'S SEIBERSDORF AGRICULTURAL LABORATORY

The Seibersdorf Laboratory has an Entomology Unit with four professional positions. These scientists conduct the R&D needed to adapt technologies for use in developing countries. In recent years, they have focused on the medfly, tsetse flies, and the New World screwworm. Their role is to fill identified technology gaps while conducting Technical Cooperation Projects and Coordinated Research Programs. They may undertake exploratory research to a limited extent, to ascertain whether certain approaches may solve a need relevant to a program in one or more developing countries. Over the years, the Seibersdorf Laboratory has made important contributions to mass rearing technology for the medfly.[102-113] They have also taken a leading role in developing genetic sexing strains of the Mediterranean fruit fly;[114-121] determining their quality, and working out the protocols for the management and mass rearing,[116,122-127] handling, sterilization, and ascertaining the field performance of such specialized strains.[34,67] Also, they have undertaken R&D to determine whether *Bacillus thuringiensis* strains may be developed as replacements for malathion in bait sprays.[128-129] The technical staff is very helpful in facilitating the procurement or fabrication of items needed in developing countries and facilitating the shipment of such items. In addition, the Seibersdorf staff serves as a "think tank" to assist the Joint FAO/IAEA Division to review and adjust programs and to chart new directions. Finally, the Seibersdorf staff plays an essential role in training scientists from developing countries. Fellowships are awarded for 3 to 12 months of training at Seibersdorf. Group training may be sponsored. Beginning in 1993, a biennial Interregional Training Course on selected applications of nuclear techniques in Entomology will be taught. The Joint Division has the lead role in identifying research priorities for the Laboratory.

1. **Research Contract Programs**

Most Research Contracts support Coordinated Research Programs (CRPs). Research Contracts are awarded based on merit directly to the institutions that employ outstanding scientists from developing countries. Through the Contract, the IAEA provides scientists in developing countries with "seed money" to defray some of the costs of conducting research for up to five years. In addition, as part of each Coordinated Research Program, unfunded Research Agreements are awarded to leading scientists from developed countries. All of these scientists are brought together about every 18 months at Agency expense to present their findings and to plan further research at a Research Coordination Meeting (RCM). In addition, Technical Contracts may be awarded one year at a time to scientists in either developing or developed countries to meet certain needs defined by the Joint Division. Applications for Research Contracts and Technical Contracts must be made using IAEA Form N-17 (Research Contract Proposal); whereas Proposals for Research Agreements must be made using IAEA Form N-20 (Proposal for Research Agreement).

2. **Technical Cooperation Projects**

Technical Cooperation "TC" Projects are administered by the Agency's Department of Technical Cooperation, but the scientific and technical leadership is provided by Technical Officers in the Joint FAO/IAEA Division. Technical Cooperation Projects are implemented primarily by providing: (1) experts to assist in the Project, (2) fellowship training, (3) scientific visits, (4) group training, (5) equipment and supplies, (6) funded subcontracts for special items (e.g. software) or services.

Applications for Technical Cooperation Projects must be made using IAEA Form TA-1 (Request for Assistance Under the IAEA Regular Program of Technical Cooperation). Technical Cooperation Projects that are approved by the Agency's Board of Governors are either funded from the Agency's Regular Budget or they are assigned to the Footnote A category for voluntary funding by a donor.

3. **Training Courses**

Every other year, the IAEA/FAO Interregional Training Course on The Use of Radiation and Isotopes on Insect Control and Entomology is held at Gainesville, Florida, U.S.A. The cost of this course is shared by the U.S. Government and the IAEA Department of Technical Cooperation's Training Courses Section. The Joint Division supplies the Technical Officer for this training course, and the Course Director is provided by the University of Florida. This Training Course is jointly funded by the United States Government and the IAEA. In addition, regional and national training courses are sponsored. For example in November 1990, a training course on the sterile insect technique and F-1 sterility for the Asia/Pacific Region was organized so that three weeks of lectures, laboratory exercises, and a field trip occurred in Malaysia. The final week was spent visiting the melon fly mass production facility on Okinawa and the field work in the Yaeyama Islands. Currently a National Training course on fruit flies is under consideration for Thailand in 1993.

4.   Information Exchange

The Joint FAO/IAEA Division organizes symposia, seminars, and other scientific meetings and publishes the proceedings. A semiannual Newsletter, describing the research accomplishments of the Seibersdorf Laboratory, is issued by the Insect and Pest Control Section, and extensive correspondence is conducted with scientists and officials around the world.

5.   Consultant Group Meetings

These are held to obtain guidance from outside experts on adjusting current Coordinated Research Programs and the program of the Entomology Unit of the Seibersdorf Laboratory and to chart the course of future programs of research, training and technical cooperation.

6.   Expert Group Meetings

These meetings are convened to obtain guidance from outside experts on technical issues relevant to specific Technical Cooperation Projects.

It is important to understand that these various mechanisms are not employed independently of each other, but they are used together in a mutually reinforcing manner to advance the Projects that have been approved both by the FAO and the IAEA and that are published in the IAEA "blue book," i.e. "The Agency's Program and Budget." For example, for 1993 - 1994, the Agency will conduct the Project on fruit flies shown in Table 2.

Many people do not clearly understand the distinctions and similarities between the Research Contract Program and the Technical Cooperation Program. These distinctions are given in Table 3, which was developed by Teresa Benson-Wiltschegg, Head, Research Contract Section.

Technical Cooperation Projects pertaining to the use of SIT against fruit flies that have been approved by the Board of Governors and that are either underway or awaiting a donor to provide funding are shown in Table 4. Proposals that will be submitted for consideration during the 1993-1994 budget cycle are shown in Table 5. It seems likely that several of these proposals, but certainly not all, will be approved.

Prior to the mid-1960s Research Contracts were awarded on an ad hoc basis. Subsequently, most Research Contracts have been awarded within the context of Coordinated Research Programs.[13] Those that have advanced the use of SIT against fruit flies are shown in Table 6.

Table 2
Project D.4.01. Improved Economy and Reliability of the Sterile Insect Technique for Use Against Fruit Flies

Objective: To develop more reliable and cost-effective technologies for applying the sterile insect technique (SIT) to control tropical fruit flies.

Project duration: 1992-1998

Tasks planned for 1993-1994	Source of Funds	Completion
1. CRP on medfly genetic sexing	Regular Budget	1993
2. Development of no-bulk larval rearing system	Regular Budget	1996
3. Field evaluation of strains and supply of large numbers of flies	Extrabudgetary & Regular Budget	Continuing
4. Development of *Bacillus thuringiensis* to replace insecticide in bait sprays	Regular Budget & Extrabudgetary	1998
5. Development of more effective attractants for female trapping, annihilation and bait sprays	Extrabudgetary & Regular Budget	1998
6. Regional CRP for Latin America on use of SIT against *Anastrepha* (1992-1998)	Extrabudgetary	1998
7. Regional CRP for Asia on use of SIT against *Bactrocera (=Dacus)* (1992-1998)	Extrabudgetary	1998
8. Development of rearing systems for genetic sexing strains (requires new rearing rooms)	Extrabudgetary & Regular Budget	1996
Forecast of TC Activities Receiving Support		
9. Support for the use of SIT in Maghrebmed medfly project-Phase 2	Extrabudgetary & Regular Budget	1995?
10. 6-8 other Technical Cooperation Projects	Regular Budget	
(Two Interregional Training Courses, also one or more Regional and National Training Courses, etc.)	Regular Budget	

Note: The Seibersdorf Laboratory is involved to varying degrees in all of these activities.

Table 3
Distinctions and Similarities Between the Research Contract Program and the Technical Cooperation Program

Both programs endeavor to address the needs of Member States, yet they take approaches that are fundamentally different.

Research or Technical contracts	Technical cooperation projects
Addresses needs as recommended by FAO/IAEA Consultant Groups with Agency-initiated CRP proposals	Addresses needs as perceived by Member states in their requests for "TC" Projects
Proposals submitted directly to IAEA by investigator's institute	Proposals submitted to IAEA through official Government channels
Solutions to specific technical problems	May provide a variety of techniques needed to solve a problem in economic development
Awards based on scientific merit of proposal and ability of investigator	Geographical distribution is an additional consideration
Institute should already have facilities needed to carry out the project	Funding may be provided to assist in creating needed facilities
Awards are small (US $2-8,000), "seed money" per year for up to 5 years	Awards are larger and cover substantial portion of project cost (ca. US$ 100,000/150,000 from 1 to 5 years)
Awards may be used to partially support institute staff and procure small equipment items and supplies	Significant awards for expensive equipment and supplies
Emphasis on achieving results related to FAO/IAEA scientific program	Emphasis on assistance to Member State in accordance with its request

What does the future hold with regard to the application of the sterile insect technique against fruit flies in the commercial production of fruits and vegetables? Because the initial costs of establishing an effective SIT program are substantial, the SIT will never be implemented merely because entomologists would like this to happen. Geier[131] explained the process whereby entomological undertakings become: (1) an essential component of the fundamental strategy of an agricultural industry and implemented in accordance with a master plan, (2) become an empirical activity practiced by some and ignored by others, or (3) pass into oblivion. Geier's[131] scheme is depicted in Figure 1.

This process usually begins with a dramatic event such as the recognition of a new pest problem, a breakthrough in developing a relevant

Table 4
Technical Cooperation Projects Involving Fruit Flies and SIT that are Being Conducted or are Awaiting a Donor to Provide Funding

Project Number	Country	Title of Project	Status
CHI/5/015	Chile	Mediterranean fruit fly eradication	Underway
COS/5/012	Costa Rica	Medfly Research Laboratory	Underway
ECU/5/013	Ecuador	Control of the Fruit Fly	Nearly done
PAK/5/018	Pakistan	Sterile Insect Technique	Nearly done
RAF/5/013	Algeria, Libya, Morocco, Tunisia	Northern Africa Regional Program for Medfly Control and Eradication: Phase I-survey and economics; Phase II-pilot trials; Phase III-operations	Phase I-ending; Phase II-awaiting donors
THA/5/038	Thailand	Integrated Control of Fruit Flies	Underway
COS/5/015	Costa Rica	Provision of Irradiation Facility	Awaiting Donor
MEX/5/018	Mexico	Sterile Insect Technique to Control *Anastrepha*	Awaiting Donor
RLA/5/025	Regional: Latin America	Sterile Insect Technique for Fruit Fly Control	Awaiting Donor

Table 5
Proposals for Technical Cooperation Projects that will be Submitted for Consideration for Funding During the 1993-1994 Budget Cycle

Country	Working Title of Proposal
Argentina	ARG/5/004 Control of fruit fly using the SIT
Bangladesh	BGD/5/016 Insect pest management by genetic manipulation
Colombia	Combatting fruit flies in Columbia
Ecuador	Fruit fly control using the sterile insect technique
El Salvador	Area-wide management of the fruit fly complex
Egypt	Use of SIT against cotton insects and the medfly
Philippines	PHI/5/022 Feasibility study of integrated control of fruit flies
Thailand	THA/5/038 Integrated control of fruit flies
Tunisia	Pilot test of SIT against Mediterranean fruit fly
Regional Project: Asia and the Pacific	Use of sterile insect technique to protect fruit and vegetables from destruction by tropical fruit flies

Table 6
Coordinated Research Programs Concerning Use of SIT
Against Fruit Flies

Title of Coordinated Research Program (CRP)	Dates	Participants (Number)	Countries (Number)
1. Fruit fly eradication or control by the sterile insect technique	1968-1977	28 also Seibersdorf	21 also Seibersdorf
2. Development of sexing mechanisms in fruit flies through manipulation of radiation induced conditional lethals and other genetic means.	1981-1988	13 also Seibersdorf	11 also Seibersdorf
3. Standardization of medfly trapping for use in sterile technique programs	1986-1991	9 also Seibersdorf	9 also Seibersdorf
4. Laboratory and field evaluation of genetically altered medflies for use in sterile insect technique programs	1988-1993	12 also Seibersdorf	8 also Seibersdorf
5. Genetic engineering technology for the improvement of the sterile insect technique	1988-1993	7 also Seibersdorf	6 also Seibersdorf

technology, or a dramatic success in using a relevant technology. For example the recent eradication of the New World screwworm, *Cochliomyia hominivorax* (Coquerel), from Libya has fired the imagination of some leaders in North Africa.[132] Similarly, influential entomologists are excited by the possibility of using genetic sexing strains to avoid damage by sterile fruit fly females to the fruit. Moreover use of genetic sexing strains would not only result in savings in mass rearing, but far more importantly, their use would open the way for the practical use of traps baited with selective female attractants. This may lead to very large savings in the monitoring of the operational program. Such use of female attractants would eliminate much of the uncertainty involved in distinguishing between released and wild insects. It would eliminate the need to sort through enormous numbers of sterile males that are captured in order to determine whether wild males are present. Also, it would diminish the need to examine fruit for larvae. Currently, monitoring represents 40-45 percent of the cost of a SIT program, and the use of female lures eventually may cut this cost in half.

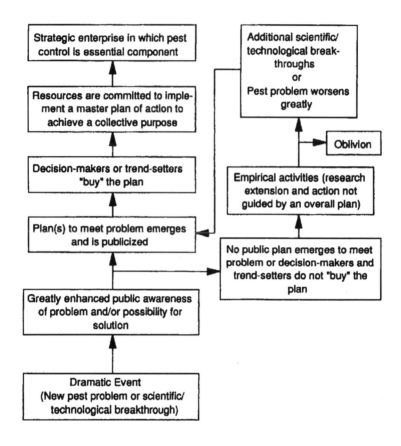

Figure 1. Process whereby pest programs become essential components of strategic enterprises, empirical activities or pass into oblivion.

The dramatic event greatly enhances public awareness of the pest problem and of the possibility of dealing more effectively with it. At this juncture, entomologists may put forward a plan to meet the pest problem in a more effective way. If the industry and political leaders accept the plan, resources in the private and public sectors can be mobilized for its implementation. If so, then the new entomological undertaking will become a fundamental component of the industry's overall strategy of producing and marketing a valuable commodity. On the other hand, if the plan is not accepted by the trend-setters in the industry, then the entomological undertaking either passes into oblivion, or it is implemented on an empirical basis. With regard to SIT, empirical implementation has occurred only against the onion maggot in the Netherlands.[133-134]

Thus, the fundamental "driver" in future implementation of the SIT against fruit flies is the perception of leaders of the fruit and vegetable industry and of political leaders. Those leaders who view the use of SIT against fruit flies as an essential component in their strategy for developing an advanced fruit and vegetable industry with well developed domestic and export markets are working for the implementation of SIT programs against tropical fruit flies. This has occurred on Japan's southern islands with regard to the melon fly and oriental fruit fly. It has occurred in Australia, Chile, Mexico and the United States, where programs have been implemented against the medfly. The United States has had an operational SIT program as a quarantine against the Mexican fruit fly along the border with Mexico since 1964, and is now beginning to use SIT against the Caribbean fruit fly in citrus growing regions of Florida in order to establish certified fly-free areas of citrus production. The latter have proven to be very important to Florida citrus growers in gaining access to the lucrative market in Japan. Moreover, Mexico has planned a twelve year program to use SIT for the removal of four species of *Anastrepha* from large fruit and vegetable production zones of Mexico. This is essential if Mexico is to increase substantially its already significant exports of fruits and vegetables. Argentina, Ecuador, Costa Rica, Guatemala, and Peru have taken significant steps in this direction. Similarly, there is strong relevant interest in the fly-free concept in the North African countries of Algeria, Libya, Morocco, and Tunisia, and some interest in El Salvador, Colombia, and the Philippines.

Other relevant assets are the large body of knowledge that has been accumulated in using SIT against fruit flies, the existence of a significant cadre of people experienced in using this technology, and significant research and development efforts in various countries and in several private firms. Thus, in considering schemes that involve the use of SIT against fruit flies paramount, attention must be given to meeting the concerns of the fruit and vegetable industries. Consequently, it is important to engage professional economists to estimate projected economic returns from the use of various alternative strategies.[135] In the case of the IAEA Technical Cooperation Project (RAF/5/013) in which solutions are sought to the medfly problem in the Maghreb region of northern Africa, an expert group calculated the net benefits based on ten-year present values for four alternative strategies.[136] They found that the use of SIT for eradication of the medfly from the Maghreb, and the use of SIT for area-wide suppression of the medfly both had substantially greater payoffs than area-wide integrated pest management without the use of SIT, or the use of SIT for the establishment and annual maintenance of certified fly free zones encompassing all orchards in the Maghreb. A second expert group asserted that establishment of various infrastructures, including quarantines that are a requirement of eradication programs, have so many additional benefits that, when feasible, eradication should be the strategy of choice.[137]

The importance of effective phytosanitary programs including quarantines can scarcely be overemphasized.[138] They are needed to safeguard the gains achieved through the use of SIT programs, or to prevent the need for such programs. By means of effective quarantines, New Zealand has been able to remain free of fruit flies, and it is the only major fruit producing nation that has done so.[139]

In conclusion, for more than three decades the Joint FAO/IAEA Division has been dedicated to advancing the use of SIT as an effective, cost-advantageous, and environmentally benign technology against tropical fruit flies. Firm foundations have been laid for significant progress in the use of this technology in the service of people around the globe.

REFERENCES

1.   Lindquist, A. W. (scientific editor), *Insect Population Control by the sterile- male technique*, Comprehensive Report of a Panel held in Vienna, 16-19 October 1962, Technical Reports Series No. 21, International Atomic Energy Agency, Vienna, 1963, 59 pp.

2.   International Atomic Energy Agency, Radiation and radioisotopes applied to insects of agricultural importance, *Proc. of a Symposium Jointly Organized by the IAEA and FAO*, Athens, 22-26 April 1963, IAEA, Vienna, 1963, 508 pp.

3.   LaBrecque, G. C. and Keller, J. C. Eds., *Advances in Insect Population Control by the Sterile-Male Technique*, Report of a Panel held in Vienna 20-24 July 1964, Technical Reports Series No. 44, IAEA, Vienna, 1965, 79 pp.

4.   Steiner, L. F. and Christenson, L. D., Potential usefulness of the sterile fly release method in fruit fly eradication programmes, *Proc. Hawaiian Acad. Sci.*, 1955-56, 17, 1956.

5.   Steiner, L. F., Harris, E. J., Mitchell, W. C., Fujimoto, M. S., and Christenson, L. D., Melon fly eradication by overflooding with sterile flies, *J. Econ. Entomol.*, 58, 519, 1965.

6.   Steiner, L. F., Mitchell W. C., and Baumhover, A. H., Progress of fruit fly control by irradiation sterilization in Hawaii and the Marianas Islands, *Int. J. Appl. Radiat. Isotopes*, 13, 427, 1962.

7.   Gilmore, J. E., Sterile insect technique: overview, in *Fruit Flies: Their Biology, Natural Enemies and Control*, Volume 3B, Robinson, A.S. and Hooper, G., Eds., Elsevier, Amsterdam, 1989.

8.   Monro, J., The Queensland fruit fly. Report given before Panel on Insect Population Control by the Sterile Male Technique, International Atomic Energy Agency. Vienna, 16-19 October 1962, 1962.

9.   Hooper, G. H. S., The sterile insect release method for suppression or eradication of fruit fly populations, in *Economic Fruit Flies of the South Pacific Region*, Plant Quar. Dept. Health, Canberra, A.C.T. Austral., 1978, 95.

10.   Steiner, L. F., Mitchell, W. C., Harris, E. J., Kozuma, T. T., and Fujimoto, M. S., Oriental fruit fly eradication by male annihilation, *J. Econ. Entomol.*, 58, 961, 1965.

11.   Steiner, L. F., Hart, W. G., Harris, E. J., Cunningham, R.T., Ohinata, K., and Kamakahi, D. C., Eradication of the oriental fruit fly from the Marianas Islands by the methods of male annihilation and sterile insect release, *Jour. Econ. Entomol.*, 63, 131, 1970.

12. **Shaw, J. G.,** A review of research done with the Mexican fruit fly and the citrus blackfly in Mexico by the Entomology Research Division, *Bull. Entomol. Soc. Amer.* 16, 186, 1970.

13. **Burditt, A. K. and Harris, E. J.,** Tropical fruit flies, in *An Evaluation of ARS Program on Alternative Methods of Insect Control,* Klassen, W., Ed., Agricultural Research Service, USDA, Beltsville, Maryland, 1976.

14. **Knipling, E. F.,** *The Basic Principles of Insect Population Suppression and Management,* Agriculture Handbook No. 512, United States Department of Agriculture, Washington, D.C., 1979, 659 pp.

15. **Klassen, W.,** Eradication of introduced arthropod pests: theory and historical practice, *Miscellaneous Publications Entomol. Soc. Amer.,* 73, 1, 1989.

16. **Monro, J.,** Experimental control of *Dacus tryoni,* in *Advances in Insect Population Control by the Sterile-Male Technique,* Report of a Panel held in Vienna, 20-24 July 1964, Technical Report Series No. 44., La Breque, G. C. and Keller, J. C., Eds., IAEA, Vienna, 1967, 22.

17. **LaChance, L. E., Schmidt, C. H., and Bushland, R. C.,** Radiation induced sterilization, in *Pest Control: Biological, Physical and Selected Chemical Methods,* Kilgore, W. W. and Doutt, R. L., Eds., Academic Press Inc., NY, 1967.

18. **Duron Aviles, E.,** Review of work to combat the Mediterranean fruit fly carried out in Central America and Panama, in *Proc. of a Panel on The Sterile-Insect Technique and its Field Applications,* 13-17 November 1972, Vienna, International Atomic Energy Agency, Vienna, 1974, 17.

19. **Turica, A.,** Status and prospects of Mediterranean fruit fly control investigations by the sterile-insect technique in Argentina, in *Proc. of a Panel on Controlling Fruit Flies by the Sterile Insect Technique,* 12-16 November 1973, Vienna, International Atomic Energy Agency, Vienna, 1975, 117.

20. **Turica, A.,** Mosca de la fruta: progresos logrados en la utilizacion de la T.I.E. en el control de la mosca del Mediterraneo en la Rep. Argentina, Seminario sobre Erradicacion de plagas mediante la tecnica del macho esteril. Lima, Peru, 1980. Publicado por la O.E.A.-CIEN, Washington, D.C., 1980, SG/Ser.Q/II-I, 1980, 13.

21. **Turica, A., Benavent, J. M., and Fernandez Vega, B.,** Segunda prueba piloto de control de la mosca del Mediterraneo mediante la liberacion de insectos esterilizados, como parte de control integrado, IDIA No. 349-354, 1977, INTA, Bs. Aires, 1979, 18.

22. **Turica, A.,** Letter to D. A. Lindquist, International Atomic Energy Agency, Vienna, 1983, 2 pp.

23. **Perdomo, A. and Linares, F.,** Evaluation of plans and projects for eradication of the medfly in Argentina, International Atomic Energy Agency, Vienna, 1990.

24. **Alvarado, L. and Ritacco, M.,** Programa Nacional de Manejo de Tefritidos (Moscas De Los Frutos), INTA y CNEA, Buenos Aires, 1991, 79 pp.

25. **Nadel, D. J and Guerrieri, G.,** Experiments on Mediterranean fruit fly control with the sterile-male technique, in *Proc. of a Panel on Sterile-Male Technique for Eradication or Control of Harmful Insects,* 27-31 May 1968, Vienna, International Atomic Energy Agency, Vienna, 1969, 97.

26. **Olalquiaga, G. F., Bobadilla T., R., Dell'Orto T., H, Ramirez H., L., Santa Cruz F., S., and Miranda O., T.,** La mosca del Mediterraneo en Chile. Ministerio de Agricultura, Direccion de Agricultura y Pesca, *Bol. Tecnico,* 20, 1, 1966.

27. **Olalquiaga, G. F.,** La lucha contra la mosca del Mediterraneo en Chile, Ministerio de Agricultura, *S.A.G. Boletin Tecnico,* 32, 1, 1967.

28. **Olalquiaga, G. F.,** Eradicacion de la mosca del Mediterraneo en la provincia de Los Andes. S.A.G. V Region de Valpraiso, 1, 1969.

29. **Harris, E. J.,** The sterile-insect technique for the control of fruit flies: a survey, in *Proc. of a Panel on Controlling Fruit Flies by the Sterile Insect Technique,* 12-16 November, 1973, Vienna, International Atomic Energy Agency, Vienna, 1975, 3.

30.   **Koyama, J.,** Second phase of the project to eradicate the Mediterranean fruit fly from Chile: Consultant Report, International Atomic Energy Agency, Vienna, Unpublished report, 1981, 27 pp.

31.   **Lopez, J. A.,** Nordenflycht, and Olalquiaga, G., *Proyecto de erradicacion y exclusion de las moscas de la fruta en Chile,* Mimeografiado, 1980, 113 pp.

32.   **Tween, G.,** Trip report: Mediterranean fruit fly in Chile, International Atomic Energy Agency, Vienna, 1990, 60 pp.

33.   **Serghiou, C., Zyngas, J., and Krambias, A.,** Preliminary studies of Mediterranean fruit fly, *Ceratitis capitata* Wied., populations in Cyprus, in *Proc. of a Panel on Computer Models and Application of the Sterile-Male Technique,* 13-17 December 1971, Vienna, International Atomic Energy Agency, Vienna, 1973, 165.

34.   **Cirio, U., Capparella, M., and Economopoulos, A. P.,** Control of medfly (*Ceratitis capitata* Wied.) by releasing a mass-reared genetic sexing strain, in *Fruit Flies: Proceedings of the Second International Symposium,* Economopoulos, A. P., Ed., 16-21 September 1986, Colymbari, Crete, Greece, Elsevier Science Pub. Co., Inc., Amsterdam, 1987, 515.

35.   **Cirio, U. and de Murtas, I.,** Status of the Mediterranean fruit fly control by the sterile-male technique on the island of Procida, in *Proc. of a Panel on The Sterile insect Technique and its Field Applications,* 13-17 Nov., 1972, Vienna, International Atomic Energy Agency, Vienna, 1974, 5.

36.   **Cirio, U.,** The Procida medfly pilot experiment: status of the medfly control after two years of sterile-insect releases, in *Proc. of a Panel on Controlling Fruit Flies by the Sterile Insect Technique,* 12-16 Nov.1973, Vienna, International Atomic Energy Agency, Vienna, 1975, 521.

37.   **Mellado, L., Caballero, R., Arroyo, M., and Jimenez, A.,** Tests on the eradication of *Ceratitis capitata* Wied. by the sterile-male technique on the Island of Tenerife (in Spanish), *Estacion de Fitopatologia Agricola,* Bol. 399, INIA, Madrid, 1966.

38.   **Mellado, L.,** La technica de machos esteriles en el control de la mosca del Mediterraneo: Programas realizados en Espana, in *Proc. of a Panel on Sterility Principle for Insect Control,* Athens, Greece, 14-18 Sept., 1970, International Atomic Energy Agency, Vienna, 1971, 49.

39.   **Mellado, L., Nadel, D. J., Arroyo, M., and Jimenez, A.,** Mediterranean fruit fly suppression experiment on the Spanish mainland in 1969, in *Proc. of a Panel on Sterile-Male Technique for Control of Fruit Flies,* 1-5 September, 1969, Vienna, International Atomic Energy Agency, Vienna, 1970, 91.

40.   **Mellado, L., Arroyo, M., and Ros, P.,** Control of *Ceratitis capitata* Wied. by the sterile-male technique in Spain, in *Proc. of a Panel on the Sterile-Insect technique and its Applications,* 13-17 November, 1972, Vienna, International Atomic Energy Agency, Vienna, 1974, 63.

41.   **Rhode, R. H.,** Application of the sterile-male technique in Mediterranean fruit fly suppression: A follow-up experiment in Nicaragua, in *Proc. of a Panel on Sterile Male Technique for Control of Fruit Flies,* 1-5 Sept., 1969, Vienna, International Atomic Energy Agency, Vienna, 1970, 43.

42.   **Rhode, R. H., Simon, J., Perdomo, A., Gutierrez, J., Dowling, Jr., C. F., and Lindquist, D. A.,** Application of the sterile-insect technique in Mediterranean fruit fly suppression, *J. Econ. Entomol.,* 64, 708, 1971.

43.   **Rhode, R. H.,** The medfly: containment in Central America, *Citrograph,* 61, 153, 1976.

44.   **Simon, J. E.,** Present status of the Peruvian project on Mediterranean fruit fly control by the release of the sterile-male technique, in *Proc. of a Panel on the Sterile Insect Technique and its Field Applications,* 13-17 November, 1972, Vienna, International Atomic Energy Agency, Vienna, 1974, 95.

45.   **Anonymous,** Final Report: Peru Medfly Project, TCP/PER/6651, June 1986 - June 1987, Unpublished manuscript, 1987, 32 pp.

46. **Perdomo Ehlers, A.,** Progress Report: Peruvian Medfly Project, PER/86/017, International Atomic Energy Agency, Vienna, 1988, 36 pp.

47. **Perdomo Ehlers, A.,** Erradicaion de la mosca del Mediterraneo en el sur del Peru, IAEA/UNDP-PER/86/017-01 Informe Tecnico 1, UNDP and IAEA, Vienna, 1989, 13 pp.

48. **Cheik, M., Howell, J. F., Harris, E. J., Ben Salah, H., and Soria, F.,** Suppression of the Mediterranean fruit fly in Tunisia with released sterile males, *J. Econ. Entomol.,* 68, 237, 1975.

49. **Kamburov, S. S., Yawetz, A., and Nadel, D. J.,** Application of the sterile-insect technique for control of Mediterranean fruit fly in Israel under field conditions, in *Proc. of a Panel on Controlling Fruit Flies by the Sterile-Insect Technique,* 12-16 Nov., 1973, Vienna, International Atomic Energy Agency, Vienna, 1975, 67.

50. **Harris, E. J., Cunningham, R. T., Tanaka, N., and Ohinata, K.,** Progress of the pilot test at Lanai against Mediterranean fruit flies and melon flies, in *Proc. of a Panel on Controlling Fruit Flies by the Sterile Insect Technique,* 12-16 November, 1973, Vienna, International Atomic Energy Agency, Vienna, 1975, 33.

51. **Harris, E. J., Cunningham, R. T., Tanaka, N., and Schroeder, W. J.,** Development of the sterile-insect technique on the island of Lanai, Hawaii for suppression of the Mediterranean fruit fly, *Proc. Hawaiian Entomol. Soc.,* 26, 77, 1986.

52. **Dreistadt, S.,** Sociopolitical impact: environmental concerns, in *Medfly and the Aftermath Symposium,* Entomol. Soc. Amer. Dec., 1982, USDA, APHIS 81-60, 1983, 36.

53. **Scribner, J.,** A review of the California action program, in *Medfly and the Aftermath Symposium,* Entomol. Soc. Amer. Dec., 1982. USDA, APHIS 81-60, 1983, 1.

54. **Ortiz, G., Liedo, P., Schwarz, A., Villasenor, A., Reyes, J., and Mata, R.,** Mediterranean fruit fly (*Ceratitis capitata*): status of the eradication programme in southern Mexico and Guatemala, in *Fruit Flies: Proceedings of the Second International Symposium,* 16-21 September 1986, Economopoulos, A. P., Ed., Colymbari, Crete, Greece, Elsevier Science Pub. Co., Inc. Amsterdam, 1987, 523.

55. **Hendrichs, J., Ortiz, G., Liedo, P., and Schwarz, A.,** Six years of successful medfly program in Mexico and Guatemala, in *Proc. of CEC/IOBC International Symposium on Fruit Flies of Economic Importance,* 16-19 Nov., 1982, Economopoulos, A. P. Ed., Athens, Greece, 1983, 353.

56. **Schwarz, A. J., Zambada, A., Orozco, D. H. S., Zavala, J. L., Calkins, C.,** Mass production of the Mediterranean fruit fly at Metapa, Mexico, *Florida Entomologist,* 68, 467, 1985.

57. **Schwarz, A. J., Liedo, J. P. and Hendrichs, J. P.,** Current programme in Mexico, in *Fruit Flies, Their Biology, Natural Enemies and Control. World Crop Pests,* Vol. 3B, Robinson, A. S. and Hooper, G., Eds., Elsevier Science Publ., Amsterdam, 1989, 375.

58. **Hentze, F. and Mata, R.,** Mediterranean fruit fly eradication programme in Guatemala, in *Fruit Flies: Proceedings of the Second International Symposium,* 16-21 September 1986, Colymbari, Crete, Greece, Economopoulos, A. P., Ed., Elsevier Science Publ., Amsterdam, 1987, 533.

59. **Fisher, K. T., Hill, A. R., and Sproul, A. N.,** Eradication of *Ceratitis capitata* (Wiedemann) (Diptera: Tephritidae) in Carnarvon, Western Australia, *J. Austral. Entomol. Soc.,* 24, 207, 1985.

60. **Foster, G. G.,** Australian genetic and sterile-insect (SIT) control programs and opportunities for collaborative research and development, Paper read at Scientific Forum on Application of Radioisotopes and Radiation to Agriculture, 10-11 Dec., 1991, Tokyo, 1991, 6 pp.

61. **Rigney, C. J.,** Sterile insect technique (SIT) control program for fruit flies of significance in Australia, Paper presented at Scientific Forum on Application of Radioisotopes and Radiation to Agriculture, Tokyo, 10-11 Dec., 1991, 3 pp.

62.   **Hendrichs, J.,** Final report on the project "Eradication of the Mediterranean fruit fly from Egypt utilizing the sterile insect technique (MISR-MED)", EGY/5/013, International Atomic Energy Agency, Vienna, 1986, 365 pp.

63.   **Cunningham, R. T.,** personal communication.

64.   **Brazzel, J. R., Calkins, C. O., Chambers, D. L., and Gates, D. B.,** Required quality control tests, quality specifications, and shipping procedures for laboratory produced Mediterranean fruit flies for sterile insect control programs, USDA-APHIS-PPQ 81-51, 3 September, 1986, Hyattsville, Maryland, 1986.

65.   **Yousef, M. Ben,** personal communication.

66.   **Arocha Rodriguez, P. and Miralles Ciscar, F.,** La mosca de la fruta en la Provincia de Santa Cruz de Tenerife. *XOBA* 2, 92, 1978.

67.   **Nitzan, Y., Rossler, Y., and Economopoulos, A. P.,** Field testing of "genetic sexing strain" for all-male release in SIT projects, Report given to Research Coordination Meeting, 24-28 September 1990, Vienna, International Atomic Energy Agency, Vienna, 1990, 11 pp.

68.   **Schwalbe, C. P.,** personal communication.

69.   **Hammes, C.,** Projet de lutte contre la mouche du Natal *Pterandrus rosa* (Karsch), Diptera Trypetidae à l'île Maurice, Rapport final, Département de l'Agriculture, des Ressources Naturelles et de l'Environnement, 1982.

70.   **Chambers, D. L., Spencer, N. R., Tanaka, N. and Cunningham, R. T.,** Sterile-insect technique for eradication or control of the melon fly and oriental fruit fly, in *Proc. of a Panel on Sterile-Male Technique for Control of Fruit Flies,* 1-5 September, 1969, Vienna, International Atomic Energy Agency, Vienna, 1970, 99.

71.   **Koyama, J.,** The Japan and Taiwan projects on the control and/or eradication of fruit flies, in *Proc. of a Symposium on Sterile Insect Technique and Radiation in Insect Control,* 29 June-3 July 1981, Neuherberg, Germany, International Atomic Energy Agency, Vienna, 1982, 39.

72.   **Koyama, J., Teruya, T., and Tanaka, K.,** Eradication of the oriental fruit fly (Diptera: Tephritidae) from the Okinawa Islands by a male-annihilation method, *J. Econ. Entomol.,* 77, 468, 1984.

73.   **Habu, N., Iga, M., and Numazawa, K.,** An eradication program of the oriental fruit fly, *Dacus dorsalis* Hendel (Diptera: Tephritidae), in Ogasawara (Bonin) Islands. 1. Eradication field test using a sterile fly release method on small islets, *Applied Entomology and Zoology,* 9, 1, 1984.

74.   **Shiga, M.,** Current programme in Japan, in *Fruit Flies: Their Biology, Natural Enemies and Control,* Volume 3B, Robinson, A. S. and Hooper, G., Eds., Elsevier, Amsterdam, 1989, 365.

75.   **Cunningham, R. T.,** Male annihilation, in *Fruit Flies: Their Biology, Natural Enemies and Control,* Volume 3B, Robinson, A. S. and Hooper, G., Ed., Elsevier, Amsterdam, 1989, 345.

76.   **Cheng, C. C. and Lee, W. Y.,** Fruit flies in Taiwan, in *Proc. of First International Symposium on Fruit Flies in the Tropics,* 14-16 March 1988, Kuala Lumpur., Malaysian Agric. Res. and Dev. Institute and Malaysian Plant Protection Society, Kuala Lumpur, 1991, 152.

77.   **Sutantawong, M.,** The present status of the sterile insect technique in Thailand, Paper read at Scientific Forum on Application of Radioisotopes and Radiation to Agriculture, 10-11 Dec., 1991, Tokyo, 1991, 11 pp.

78.   **Hendrichs, J.,** Travel report: Thailand. THA/5/038 Integrated control of fruit flies, 8-22 June 1991, International Atomic Energy Agency, Vienna, 1991, 30 pp.

79.   **Manoto, E. C. and Resilva, S. S.,** Integrated control of oriental fruit fly based on SIT, *Phil. Technology. J.,* 14, 1, 1989.

80. **Manoto, E. C.,** Status of sterile insect technique in the Philippines, Paper read at the Scientific Forum on Application of Radioisotopes and Radiation to Agriculture, 10-11 December 1991, Tokyo, Japan, 1991, 6 pp.

81. **Sutantawong, M.,** Control of fruit flies, *Dacus dorsalis* Hendel and *Dacus correctus* Bezzi by sterile insect technique at Ang Khang, Chiang Mai, Report of the Office of Atomic Energy for Peace, 1988 (in Thai).

82. **Meksongsee, B., Liewanich, A., and Jirasuratana, M.,** Fruit flies in Thailand, in *Proc. of First International Symposium on Fruit Flies in the Tropics,* 14-16 March 1988, Kuala Lumpur, Malaysian Agric. Res. and Dev. Institute and Malaysian Plant Protection Society, Kuala Lumpur, 1991, 83.

83. **Iwahashi, O.,** Eradication of the melon fly, *Dacus cucurbitae* from Kume Is., Okinawa with the sterile insect release method, *Research on Population Ecology,* 19, 87, 1977.

84. **Koyama, J.,** Eradication of the fruit flies-Kume Island to Okinawa Islands: eradication commemorative lecture, 7 Nov., 1990, Okinawa Prefectural Government, Naha, Japan, 1990, 2 pp. (in Japanese).

85. **Kakinohana, H., Kuba, H.,Yamashita, K., and Taniguchi, M.,** Eradication of the melon fly, Dacus cucurbitae, from Miyako Islands, Okinawa, with the sterile insect technique, in *Proc. of First International Symposium on Fruit Flies in the Tropics,* 14-16 March 1988, Kuala Lumpur, Malaysian Agric. Res. and Dev. Institute and Malysian Plant Protection Society, Kuala Lumpur, 1991, 232.

86. **Yoshizawa, O.,** Current state of the Japan projects for the eradication of pest insects by sterile insect technique (SIT): eradication campaign of melon flies in southwestern islands, Paper read at Scientific Forum on Application of Radioisotopes and Radiation to Agriculture, 10-11 Dec., 1991, Tokyo, 1991, 5 pp.

87. **Monro, J. and Osborn, A. W.,** The use of sterile males to control populations of the Queensland fruit fly, *Dacus tryoni* (Frogg.) (Diptera: Tephritidae). I. Methods of mass-rearing, transporting, irradiating and releasing sterile flies, *Austral. J. Zoology,* 15, 461, 1967.

88. **Melis, A. and Baccetti, B.,** Metodi di lotta vecchi e nuovi sperimentati contro principale fitofagi dell' olivo in Toscana, *Estratto da Redia,* XLV, 193, 1960.

89. **Economopoulos, A. P., Avtzis, N., Zervas, G., Tsitsipis, J., Haniotakis, G., Tsiropoulos, G., and Manoukas, A. G.,** Control of the olive fruit fly, *Dacus oleae* (Gmelin) by combined effect of insecticides and release of gamma-sterilized insects, *Z. angew. Entomol.,* 83, 201, 1977.

90. **Wang, H. S., Zhao, C. D., Li, H. X., Lou, H. Z.,Liu, Q. R., Kang, W., Hu, J. G., Zhang, H. Q., Chu, J. M., Xia, D. R., and Yang, R. X.,** Control of *Dacus citri* by irradiated male sterile technique, *Acta Agriculturae Nucleatae Sinica,* 4, 135, 1990.

91. **Kang Wen,** Status of SIT and F-1 sterility in China and proposal of operation, Paper presented at Scientific Forum on Application of Radioisotopes and Radiation to Agriculture, Tokyo, 10-11 Dec., 1991, 8 pp.

92. **Qureshi, Z. A.,** The ability of sterilized irradiated males of *Dacus cucurbitae* Coq. and *Dacus zonatus* (Saunders) to compete with normal males, in *Proc. of a Panel on Insect Ecology and the Sterile-Male Technique,* 7-11 August 1967, Vienna, International Atomic Energy Agency, Vienna, 1969, 43.

93. **Butt, B. A.,** Sterile insect technique: report to Government of Pakistan, IAEA-TA-2214, International Atomic Energy Agency, Vienna, 1984, 8 pp.

94. **Christenson, L. D.,** Radiation sterilization of the Mexican, oriental and melon flies, Report given before Panel on Insect Population Control by the Sterile Insect Technique, 16-19 October 1962, Vienna, International Atomic Energy Agency, Vienna, 1962.

95. **Lopez D. F.,** Sterile-male technique for eradication of the Mexican and Caribbean fruit flies: review of current status, in *Proc. of a Panel on Sterile-Male Technique for the Control of Fruit Flies,* 1-5 Sept., 1969, Vienna, International Atomic Energy Agency, Vienna, 1970, 111.

96.   **Zambada Martinez, A.,** Aplicacion del concepto de zona libre de moscas de la fruta en el estado de Sonora, in *IV Curso Internacional de Capacitacion en Moscas de la Fruta, II(IV)*, Secretaria de Agricultura y Recursos Hidraulicos, Mexico City, 1990, 111.

97.   **Mexican Secretariat of Agriculture and Water Resources,** Fruit flies eradication campaign (through the use of integrated pest control for sanitary and improved Mexican fruit growing), 12 year plan: executive summary, campaign document and six annexes, 1991.

98.   **Burditt, A. K., Jr., Lopez D., F., Steiner, L. F., Windeguth, D. L., Baranowski, R. M., and Anwar, M.,** Application of sterilization techniques to *Anastrepha suspensa* Loew in Florida, U.S.A., in *Proc. of a Symp. on Sterility Principle for Insect Control*, 22-26 July 1974, Innsbruck, International Atomic Energy Agency, Vienna, 1975, 93.

99.   **Holler, T. C. and Harris, D. L.,** Efficacy of sterile releases of Caribbean fruit fly adult, *Anastrepha suspensa* (Tephritidae) against feral populations in urban hosts adjacent to commercial citrus, in *Fruit Flies: Biology and Management*, Aluja, M. and Liedo, P., Eds., Springer-Verlag, New York, 1993, 329.

100.   **Boller, E. F. and Remund, U.,** Application of SIT on the European cherry fruit fly, *Rhagoletis cerasi* L., in northwest Switzerland, in *Proc. of a Panel on Controlling Fruit Flies by the Sterile Insect Technique*, 12-16 November 1973, International Atomic Energy Agency, Vienna, 1975, 77.

101.   **Boller, E. F. and Remund, U.,** Field feasibility study for the application of SIT in *Rhagoletis cerasi* L. in northwest Switzerland, in *Fruit Flies of Economic Importance*. Cavalloro, R., Ed., Proc. CEC/IOBC Internat. Symp., Athens, November, 1982, Balkema, Rotterdam, 1983, 366.

102.   **Bruzzone, N. D.,** Recycling larval media for mass-rearing the Mediterranean fruit fly, in *Fruit Flies: Proceedings of the Second International Symposium*, Economopoulos, A. P., Ed., 16-21 September 1986, Colymbari, Crete, Greece, Elsevier Science Pub. Co., Inc., Amsterdam, 1987, 277.

103.   **Bruzzone, N. D. and Schwarz, A. J.,** Recycling larval rearing medium for Mediterranean fruit fly mass-production: a preliminary experiment, *J. Appl. Entomol.*, 103, 418, 1987.

104.   **Economopoulos, A. P., Bruzzone, N. D., Judt, S., Wornoayporn, V., El-Agal, M. N., Al-Taweel, A. A., and Wang, H. S.,** Recent developments in medfly mass rearing, in *Proc. of a Symp. on Modern Insect Control: Nuclear Techniques and Biotechnology*, 16-20 November, 1987. Vienna, International Atomic Energy Agency, Vienna, 1988, 285.

105.   **Fay, H. A. C.,** A starter diet for mass-rearing larvae of the Mediterranean fruit fly, *Ceratitis capitata* (Wied.), *J. Appl. Entomol.*, 105, 496, 1988.

106.   **Economopoulos, A. P. and Bruzzone, N. D.,** Mass rearing of the Mediterranean fruit fly, (Diptera: Tephritidae): continuous light in the adult stage, *J. Econ. Entomol.*, 82, 1492, 1989.

107.   **Economopoulos, A. P., Bruzzone, N. D., and Judt, S.,** Effect of continuous light and oviposition hole size on the egg production of Mediterranean fruit flies mass reared for sterile insect technique programmes in *Regulation of Insect Reproduction IV*, (Proc. Symp., Zinkovy, 1987), Academia Publishing House, Czechoslovak Academy of Sciences, Praha, 1989, 193.

108.   **Economopoulos, A. R. and Judt, S.,** Effect of oviposition net hole size and treatment of net with sugar or lubricant-release agent on medfly egg production and collection, in *Fruit Flies: Proceedings of the Second International Symposium*, Economopoulos, A. P., Ed., 16-21 September 1986, Colymbari, Crete, Greece, Elsevier Science Pub. Co., Inc., Amsterdam, 1987, 325.

109.   **Galun, R.,** Phagostimulation of the Mediterranean fruit fly, *Ceratitis capitata* by ribonucleotides and related compounds, *Entomol. Exp. & Appl.*, 50, 133, 1989.

110. **Hooper, G. H. S.,** Effect of pupation environment on the quality of pupae and adults of the Mediterranean fruit fly, *Entomol. Exp. & Appl.,* 44, 155, 1987.

111. **Hooper, G. H. S.,** Application of quality control procedures to large-scale rearing of the Mediterranean fruit fly, *Entomol. Exp. & Appl.,* 44, 161, 1987.

112. **Monro, J.,** Improvements in mass rearing the Mediterranean fruit fly, in *Proc. of a Panel on Radiation, Radioisotopes and Rearing Methods in the Control of Insect Pests,* 17-21 October 1966, Tel Aviv, International Atomic Energy Agency, Vienna, 1968, 13.

113. **Nadel, D. J.,** Current mass-rearing techniques for the Mediterranean fruit fly, in *Proc. of a Panel on Sterile Male Technique for Control of Fruit Flies,* 1-5 Sept., 1969, Vienna, International Atomic Energy Agency, Vienna, 1970, 13.

114. **Busch-Petersen, E.,** Progress in the isolation of temperature-sensitive lethal factors for genetic sexing in *Ceratitis capitata* (Wied.), in *Proc. XVII International Congress of Entomology,* September, 1983, Hamburg, F.R.G., 1984.

115. **Busch-Petersen, E.,** Development of induced sex separation mechanisms in *Ceratitis capitata* (Wied.): EMS tolerance and suppression of female recombination, in *Fruit Flies: Proceedings of the Second International Symposium,* Economopoulos, A. P., Ed., 16-21 September 1986, Colymbari, Crete, Greece, Elsevier Science Pub. Co., Inc., Amsterdam, 1987, 209.

116. **Busch-Petersen, E. and Baumgartner, H.,** Simulation of instability in genetic sexing strains: effects of male recombination in association with other biological parameters, *Bull. Entomol. Res.,* 81, 11, 1991.

117. **Franz, G. and Busch-Petersen, E.,** Analysis of a temperature sensitive lethal mutation used as selectable marker in a genetic sexing strain of the Mediterranean fruit fly, *Ceratitis capitata,* 1992 (In preparation).

118. **Kerremans, P. and Busch-Petersen, E.,** Polytene chromosome analysis in relation to genetic sex-separation mechanisms in *Ceratitis capitata* (Wied.), in *Genetic Sexing of the Mediterranean Fruit Fly. Proc. of Final Research Coordination Meeting,* Colymbari, Crete, Greece, 3-7 September 1988, International Atomic Energy Agency, Vienna, 1990, 61.

119. **Kerremans, P. and Franz, G.,** Cytogenetic analysis of chromosome 5 of the Mediterranean fruit fly, *Ceratitis capitata,* 1993 (In press).

120. **Kerremans, P., Genscheva, E., and Franz, G.,** The sterile insect technique: improving genetic sex-separation methods for the medfly, *Ceratitis capitata, Med. Fac. Landbouww. Rijksuniv. Gent,* 56, 1053, 1991.

121. **Kerremans, P., Genscheva, E., and Franz, G.,** Genetic and cytogenetic analysis of Y-autosome translocations in the Mediterranean fruit fly, *Ceratitis capitata, Genome,* 135, 1993 (in press).

122. **Busch-Petersen, E.,** Male recombination in a genetic sexing strain of *Ceratitis capitata* (Diptera: Tephritidae), and its impact on stability, *Ann. Entomol. Soc. Amer.,* 82, 778, 1989a.

123. **Busch-Petersen, E.,** Genetic sex separation in sterile insect technique pest management programmes, with special reference to the medfly, *Ceratitis capitata* (Wied.), in *Fruit Flies of Economic Importance,* Cavalloro, R., Ed., 87. Proc. of the CEC/IOBC International Symposium, Rome, Italy, 7-10 April, 1987, A. A. Balkema, Rotterdam, 1989b, 225.

124. **Busch-Petersen, E., Ripfel, J., Pyrek, A., and Kafu, A.,** Isolation and mass rearing of a pupal genetic sexing strain of the Mediterranean fruit fly, *Ceratitis capitata* (Wied.) in *Proceedings of a Symposium on Modern Methods of Insect Control: Nuclear Techniques and Biotechnology,* Vienna, 16-20 November 1987, International Atomic Energy Agency, Vienna, 1988, 211.

125. **Busch-Petersen, E. and Kafu, A.,** Stability of two mass-reared genetic sexing strains of the Mediterranean fruit fly, *Ceratitis capitata* (Wiedemann) (Diptera: Tephritidae), based on pupal colour dimorphisms, *Environ. Entomol.,* 18, 319, 1989a.

126. **Busch-Petersen, E. and Kafu, A.,** Assessment of quality parameters during mass-rearing of a genetic sexing strain of the Mediterranean fruit fly, based on a pupal colour dimorphism, *Entomol. Exp. & Appl.,* 51, 241, 1989b.

127. **Hendrichs, J.,** Travel report: Thailand. THA/5/038 Integrated control of fruit flies, 8-22 June 1991, International Atomic Energy Agency, Vienna, 1991, 30 pp.

128. **Gingrich, R. E.,** Demonstration of *Bacillus thuringiensis* as a potential control agent for the adult Mediterranean fruit fly, *Ceratitis capitata* (Wied.), *J. Appl. Entomol.,* 104, 378, 1987.

129. **Gingrich, R. E. and El-Abbassi, T. S.,** Diversity among *Bacillus thuringiensis* active against the Mediterranean fruit fly, *Ceratitis capitata,* in *Proc. of a Symp. on Modern Insect Control: Nuclear Techniques and Biotechnology,* 16-20 November, 1987, Vienna, International Atomic Energy Agency, Vienna, 1988.

130. **Offori, E. D. and Butt, B. A.,** Programme and activities of the Insect and Pest Control Section: 1964-1989, *Joint FAO/IAEA Division of Nuclear Techniques in Food and Agriculture,* International Atomic Energy Agency, Vienna, 1989, 37 pp.

131. **Geier, P. W.,** Temporary suppression, population management, or eradication: how and when to choose, *in Concepts of Pest Management,* Rabb R. L. and Guthrie, F. E., Eds., Proc. of Conference at North Carolina State University, Raleigh, N.C., U.S.A., 25-27 March 1970, 1970, 242 pp.

132. **Lindquist, D. A., Abusowa, M., and Hall, M. J.,** The New World screwworm fly in Libya: its introduction and eradication, *Medical and Veterinary Entomology,* 1992 (In press).

133. **Loosjes, M.,** The onion fly: special report on its control in the Netherlands by the sterile insect technique, *Insect and Pest Control Newsletter,* 44, 13, 1990.

134. **Van Heemert, C. and Robinson, A.,** Ten years of translocations in the melon fly, *Delia (= Hylemya) Antigua,* in *Proc. of a Symposium on Sterile Insect Technique and Radiation in Insect Control,* 29 June-3 July, 1981, Neuherberg, International Atomic Energy Agency, Vienna, 1982, 445.

135. **Reichelderfer, K. H., Carlson, G. A., and Norton, G. A.,** Economic guidelines for crop pest control, FAO Plant Production and Protection Paper No. 58, Food and Agriculture Organization, Rome, 1984, 93 pp.

136. **Driouchi, A., Carlson, C. A., Mumford, J. D., and Enkerlin, W.,** Economic evaluation of alternative strategies for medfly control in the Maghreb, Report to International Atomic Energy Agency, Vienna, 1992.

137. **Rohwer, G. G., Tween, G., and Reyes F., J.,** A programme for the eradication of the Mediterranean fruit fly from the Maghreb, Report to the International Atomic Energy Agency, Vienna, 1992.

138. **Hedley, J.,** Improved plant quarantine using bilateral quarantine agreements for the exclusion of high risk pests, *FAO Plant Prot. Bull.,* 38, 127, 1990.

139. **Baker, R. T., Cowley, J. M., Harte, D. S., and Frampton, E. R.,** Development of a maximum pest limit for fruit flies (Diptera: Tephritidae) in produce imported into New Zealand, *J. Econ Entomol.* 83, 13, 1990.

Chapter 2

POPULATION GENETICS OF *CERATITIS CAPITATA* AND PHYLOGENETIC RELATIONS WITH OTHER TEPHRITIDAE

G. Gasperi, A. R. Malacrida, L. Baruffi, R. Milani,
and C. R. Guglielmino

I. INTRODUCTION

The Mediterranean fruit fly *Ceratitis capitata* is a polyphagous and multivoltine tropical species which in the last hundred years has spread from its supposed origin in Africa to a number of countries including the Mediterranean basin, parts of South and Central America, and Australia.[1]

Information on dispersal and on colonization rate of this species,[2] coupled with knowledge of the genetic structure of natural populations,[3] is crucial in gaining an understanding of life history strategies that determine the rate of gene flow between different sub-populations, and in planning management strategies against this pest. Intraspecific genetic differentiation is expected to be associated with the colonization history of *C. capitata*,[4-5] despite the apparent morphological uniformity within this species.[6]

From the taxonomic point of view there is no recent work on the genus *Ceratitis* and on its phylogenetic relationships with other Tephritidae flies. A genetic analysis may offer a suitable approach to clarify the systematics and phylogeny of these species.

II. MATERIALS AND METHODS

A. POPULATIONS OF *CERATITIS CAPITATA*

Two populations from the African region (Kenya and Réunion Isl.), two populations from the Mediterranean basin (Sardinia and Procida Island in the Bay of Naples, Italy), and one American population (from Guatemala) were studied according to sampling opportunities. Therefore, five geographic areas are represented: Kenya, Réunion, Procida, Sardinia, and Guatemala.

B. SPECIES AND GENUS LEVEL SYSTEMATICS

Population samples of *C. capitata*, *Ceratitis rosa*, *Trirhithrum coffeae* were collected in Kenya, and specimens of *Capparimyia savastani* were collected in Pantelleria Isl. (Italy).

C. ELECTROPHORETIC STUDIES

Preparation of samples and electrophoretic procedures are described by Gasperi et al.[4] For each population sample at least 25 individuals were assayed at the following 27 enzyme loci: Mpi, Est_6, Mdh_2, Hk_2, Est_1, Est_2,

0-8493-4854-4/94/$0.00+$.50
© 1994 by CRC Press, Inc.

Table 1
Parameters of Genetic Variability in Four Populations of *C. capitata*

Geographic origin	Year	A ± sd	P ± sd	H ± sd
KENYA		2.000 ± .16	.568 ± .08	.173 ± .01
REUNION		1.527 ± .26	.403 ± .14	.116 ± .07
SARDINIA		1.337 ± .068	.238 ± .06	.084 ± .026
GUATEMALA		1.154	.077	.039 ± .030
	1983	1.185 ± .08	.166 ± .07	.031 ± .010
	1984	1.172 ± .08	.173 ± .08	.042 ± .016
PROCIDA	1985	1.277 ± .09	.240 ± .07	.066 ± .018
	1986	GS T-101 Strain		
	1986	1.245 ± .13	.228 ± .11	.045 ± .018

Pgi, *Zw*, *Pgd*, *Fh*, *Had*, Hk_1, *Idh*, *Pgm*, Got_1, Got_2, Ak_2, Mdh_1, Adh_2, *Gpt*, *Pgk*, *Me*, α-*Gpdh*, *Aox*, Ak_1, $Acon_1$, $Acon_2$.

D.   DATA ANALYSIS

F_{ST} statistics were computed by the method of Nei and Chesser.[7] Cluster analysis through a tree representation was performed using the BIOSYS-1 program of Swofford and Selander.[8]

III. RESULTS

A.   GENETIC STRUCTURE OF POPULATIONS

The levels of genetic variability for each population are summarized in Table 1. A general trend seems to characterize these populations of *C. capitata*, which are polymorphic to different degrees. In particular, the Mediterranean populations (Procida and Sardinia) and the American one (Guatemala) are less polymorphic than subtropical (Réunion) and tropical ones (Kenya). This is supported by all three statistics A (average number of alleles per locus), P (proportion of polymorphic loci) and H (average number of herozygous individuals). These findings are in agreement with the general rule of a decreasing trend of genetic variability away from the source area of a species towards the periphery of its geographic range. The Kenya population has the highest average number of alleles per locus (2.00 ± 0.16); the major contribution to this high value is due to the presence, in this population, of several low frequency alleles. The number of rare alleles present only in this population can be estimated to be about 25% of the total number of alleles. The discovery of several low frequency alleles in Kenya supports the hypothesis that this population has maintained a large size.[9]

B.   ANNUAL ANALYSIS IN PROCIDA

The Procida samples collected from 1983 to 1985 are presented in Table 1. It is noteworthy that 1986 is the year in which the genetic sexing strain (T-101) was released in Procida: these samples represent the after-treatment rising population.[10] The average heterozygosity (H), proportion of polymorphic loci (P) and mean number of alleles per locus (A), observed for each year in Procida samples show that the mean number of recorded heterozygous individuals has doubled in a three year period (the difference is significant, $t_{(10)} = 3.3$, $0.01 > p > 0.001$). These data may indicate a trend of increasing variability in the Procida medfly population. However the relatively short period of observations and a not systematic monthly sampling suggest some cautions in the interpretation.

The mean heterozygosity value of the samples collected in 1986, after the release of a sexing strain, is similar to the one assessed in 1984. The lowering of the mean heterozygosity between the 1985 and 1986 samples is close to statistical significance ($t_{(11)} = 1.87$, $0.10 > P > 0.05$) and can be related to the release of sterile males.

C.   SINGLE-LOCUS   MEASURES   OF   POPULATION   DIFFERENTIATION

The pattern of allelic differentiation within and between geographically distant populations of $C.$ $capitata$ has been analyzed in order to assess the degree of genetic divergence in relation to ecogeographical conditions. The proportion of heterozygosity attributable to differentiation between populations is represented by the fixation index (F_{ST}). Estimates of F_{ST} for each of the polymorphic loci are partitioned according to linkage group because of the possibility that the linkage among loci might affect their pattern of variability (Table 2).

All linkage groups, except A (Chr. 4), appear to have the same value for this parameter and the average values of F_{ST} range from 0.036 to 0.094. For linkage group A, the mean value of F_{ST} is high, 0.379. In this linkage group Est_1, Est_2 and to a lesser degree Hk_2 contribute substantially to the geographic differentiation of $C.$ $capitata$. It is noteworthy that the loci Est_1 and Est_2 are tightly linked and are located at 26.8 ± 0.03 cM from Hk_2.[11] The average of all the F_{ST} values listed in Table 2 is 0.123; this estimate indicates that at least 12% of the total variability in $C.$ $capitata$ is attributable to divergence between populations. But an important feature of the results listed in Table 2 is the wide range of F_{ST} values among loci (from 0.685 at Est_2 to 0). As drift alone would affect all loci similarly, the observed F_{ST} heterogeneity may be caused by some form of selection. Part of the differentiation could simply be due to drift and part could be maintained by some sort of selection affecting particular genes.[12]

Table 2
F- Statistics Analysis for the Enzyme Loci Located on Various Chromosomes, in Geographical Populations of *C. capitata*

Locus	N. of alleles	F_{ST}	F_{ST}
	Linkage group A (Chr. 4)		
Hk_2	4	0.219	
Est_1	3	0.613	0.379
Est_2	3	0.658	
Pgi	2	0.000	
	Linkage group B (Chr. 5)		
Zw	2	0.117	
Pgd	3	0.032	
Fh	2	0.000	0.093
Had	2	0.130	
Hk_1	2	0.147	
Gpt	2	0.033	
	Linkage group C (Chr. 3)		
Got_2	2	0.183	0.094
Mdh_1	3	0.005	
	Linkage group D (Chr. 2)		
Mpi	7	0.108	
Est_6	2	0.126	
Mdh_2	2	0.004	0.059
Adh_2	2	0.000	
	Linkage group E (Chr. 6)		
Idh	3	0.006	
Pgm	5	0.069	0.036
Got_1	2	0.032	
	Unmapped loci		
Ak_2	2	0.139	
Pgk	3	0.066	0.069
Me	2	0.002	

D. GENES AND POPULATION DIFFERENTIATION

The contribution of single loci to population differentiation is presented in Figure 1. If all populations are considered, the distribution of F_{ST} is mainly around 0.14 and only four loci show values around or greater than 0.20. These loci are Est_2, Est_1, Hk_2, $Got2$. A large portion of the interpopulation differentiation is contributed by the Réunion population, as only Hk_2 shows F_{ST} values greater than 20% if this population is omitted from the computation. Réunion seems to be significantly differentiated from all the other populations at the *Est* loci, due to the presence in this population

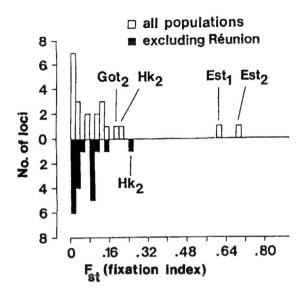

Figure 1. The distribution of the fixation index (F_{ST}) at polymorphic loci in the geographical populations of *Ceratitis capitata* considering all populations and excluding Réunion.

of fixed alleles alternative to the most common in the other populations. When the other populations are omitted from the computation, no significant changes are observed for the fixation index of the above mentioned loci.

E.   SPATIAL AND TEMPORAL DIFFERENTIATION

A tree obtained with Cavalli, Sforza, and Edwards' distances[13] representing all the 36 samples from the two native (Kenya, Réunion) and three introduced populations (Sardinia, Procida, and Guatemala) is shown in Figure 2. Out of the 36 samples analyzed, 22 belong to the Mediterranean Procida population and represent collections from different years and months. The main feature of this tree is the entangling of geographic and seasonal differentiation among the samples: Procida early season samples (April and February) appear separated by a major split, in between Réunion samples and Kenya-Guatemala samples.

F.   F STATISTIC ANALYSIS

The F statistics are used in an attempt to distinguish the various patterns of genetic variation and to infer the underlying causes and their relative contribution to the total variation.

Cavalli Sforza and Edwards' distance

P = Procida samples
* = April and February samples of Procida

Figure 2. Tree representing the 36 samples of *Ceratitis capitata*. The genetic distances are calculated according to Cavalli, Sforza, and Edwards[13] and are based on all the allelic frequencies, including those of *Mpi locus*.

In Table 2 the subpopulations represent the considered geographic areas. Among the polymorphic loci, *Mpi* didn't show differentiation at the geographic level. On the other hand, if the same data are analyzed over time between months (i.e. when the subpopulations are the single month collections) *Mpi* shows a degree of differentiation (F_{ST} = 0.320), whereas no significant F_{ST} values are found at other loci. In this analysis the numerous Procida samples had the major load.

When the Procida samples are considered separately and are compared across three seasons (Dec.-April/June-Aug./September) a very high degree of differentiation appears for the *Mpi* locus. The F_{ST} value across seasons, for this locus, amounts to 0.509, while the corresponding across year value is 0.017. In the Procida population, there is a trend in *Mpi* allele frequencies during the year. Only in the three samples (April 1983, December 1984, February 1985), representing the wintering season, the Mpi^{105} allele appears with an unexpected high frequency. The statistically significant gene heterogeneity across the seasons at the *Mpi* locus causes the entangling of spatial and temporal variation portrayed in the tree of Figure 2. In fact when this locus is removed from the cluster analysis, the seasonal pattern disappears and the tree samples split according to a geographic pattern of differentiation.[5]

G. PHYLOGENETIC RELATIONS OF *CERATITIS CAPITATA* WITH OTHER TEPHRITIDAE

The electrophoretic variability at 27 hortologous enzyme loci, has been used to study the phylogenetic relationships among species of *Ceratitis*, *Trirhithrum*, and *Capparimyia* genera.[14] The agreement between this genetic approach and the existing conventional systematics appears low in relation to the species of the genera *Ceratitis* and *Trirhithrum*. *C. capitata* and *T. coffeae* appear closer than the two *Ceratitis* species as evaluated by Nei, Roger, and Wagner distances.[14]

IV. DISCUSSION

The results obtained in this study focus attention on the pattern and causes of geographic differentiation in natural populations of *C. capitata*. In the analysis of single locus differentiation, some loci show a significant amount of genetic differentiation between populations. The population mainly involved in this process is Réunion as shown in Figure 1 because of the presence of fixed alleles at Est_1 and Est_2 loci. If the differentiation indices (F_{ST}) are considered separately in relation to each linkage group (Table 2), their estimates provide grounds to believe that linkage constrains the pattern of variability. In fact, the mean fixation index F_{ST} is about 12% and the highest contribution to this value is given by loci of linkage group A (Chr. 4). For these loci, segregation distortion, due to the presence of Sd_1 factor[3-15] may be one cause for the observed deviation from panmixia.

On the other hand, estimates of variability not involving gene frequencies directly, such as the average number of alleles per locus (A), the proportion of polymorphic loci (P), and the average number of heterozygous individuals (H) seem related to the populations' evolutionary history of the medfly. Based on these parameters, it is observed that the Réunion samples are much more similar to the Kenyan samples than to the distant Mediterranean populations and the Guatemala sample. These statistics reveal a trend of decreasing variability between flies from the putative source area (Africa) and those from the periphery of the dispersion area. The dispersion itself might explain a reduction of variability in the derived Mediterranean populations, as a consequence of a bottleneck in population size.[9] This low initial level of variability may be maintained over time because of different environmental constraints, e.g. seasonal changes. Seasonality may affect the demographic profile of a Mediterranean population. Mediterranean medfly populations experience seasonal cycle variation since winter and summer generations develop under very different conditions. The anomalous allocation of cold seasons samples in the tree of Figure 2 is due to the allelic variation at the *Mpi* locus, which shows very high F_{ST} values across seasons. The deviation from panmixia at the *Mpi* locus can hardly be attributed to inbreeding, given the absence of analogous indications for other linked loci.[5] Nevertheless an inbreeding effect cannot be completely excluded.

Therefore, different evolutionary forces could have contributed, singly or in concert, to genetic changes during colonization by *C. capitata*. Selection and/or bottleneck effect may have played an important role, in the differentiation of the Réunion samples, which on the other hand maintain the genetic attributes of an ancestral population. Drift, which is correlated with geographic distance seems to have played a major role in the dispersion processes of the Mediterranean populations.

A genetic analysis has been attempted to clarify the phylogenetic relations of *C. capitata* with some other Tephritidae species. The results obtained in this approach are in disagreement with the current morphological taxonomy because *C. capitata* and *T. coffeae* appear closer than the two *Ceratitis* species. From the morphological point of view, *Ceratitis (Ceratitis) capitata* and *Ceratitis (Pterandrus) rosa* are separated on the basis of the male secondary sexual characters, with females being inseparable at the generic level.[16] These disagreements concern taxa which had been assigned to various taxonomic positions.

ACKNOWLEDGEMENTS

This research was supported by the National Research Council of Italy, Special Project RAISA, Sub-project No. 2, Paper No. 532. Grants from I.A.E.A. (International Atomic Energy Agency), and from MPI (funds 40 per cent from National Ministry of Education) also contributed to this work.

REFERENCES

1. **Fletcher, B. S.,** Life history strategies of tephritid fruit flies, in *Fruit Flies: Their Biology, Natural Enemies and Control,* Vol. 3B, Robinson, A. S and Hooper G. H., Eds., Elsevier Science Publ., Amsterdam, 1989, 195.
2. **Fletcher, B. S.,** Movements of tephritid fruit flies, in *Fruit Flies: Their Biology, Natural Enemies and Control,* Vol. 3B, Robinson A. S. and Hooper G. H., Eds., Elsevier Science Publ., Amsterdam, 1989, 209.
3. **Milani, R., Gasperi, G., and Malacrida, A.,** Biochemical Genetics, in *Fruit Flies: Their Biology, Natural Enemies and Control,* Vol. 3B, Robinson, A. S. and Hooper G., Eds., Elsevier Science Publ., Amsterdam, 1989, 33.
4. **Gasperi, G., Guglielmino, C. R., Malacrida, A., and Milani, R.,** Genetic variability and gene flow in geographical populations of *Ceratitis capitata* (Wied.) (medfly), *Heredity* , 67, 347, 1991.
5. **Malacrida, A. R., Guglielmino, C. R., Gasperi, G., Baruffi, L., and Milani, R.,** Spatial and temporal differentiation in colonizing populations of *Ceratitis capitata,* *Heredity,* 69, 101, 1992.
6. **White, I. M.,** The state of fruit fly taxonomy and future research priorities, in *Fruit Flies of Economic Importance 87,* Cavalloro, R., Ed., A.A. Balkema, Rotterdam, 1989, 543.
7. **Nei, M. and Chesser, R. K.,** Estimation of fixation indices and gene diversities, *Ann. Hum. Genet.,* 47, 253, 1983.
8. **Swofford, D. L. and Selander, R. B.,** BIOSYS-1. A computer program for the analysis of allelic variations in genetics, Univ. Illinois, Urbana, 1981.
9. **Nei, M., Maruyama, T., and Chakraborty, R.,** The bottleneck effect and genetic variability in populations, *Evolution,* 29, 1, 1975.
10. **Cirio, U., Capparella, M., and Economopoulos, A. P.,** Control of medfly (*Ceratitis capitata* Wied.) by releasing a mass-reared genetic sexing strain, in *Fruit Flies,* Economopoulos, A. P., Ed., Elsevier Science Publ., Amsterdam, 1987, 515.
11. **Gasperi, G., Malacrida, A., Tosetti, M., and Milani, R.,** Enzyme variability: a tool for investigating the genome organization and the population structure of *Ceratitis capitata,* in *Fruit Flies of Economic Importance 84,* Cavalloro, R., Ed., Balkema, A. A., Rotterdam, 1986, 153.
12. **Singh, R. S. and Rhomberg, L. R.,** A comprehensive study of genetic variation in natural populations of *Drosophila melanogaster* . II. Estimates of heterozygosity and patterns of geographic differentiation, *Genetics,* 117, 255, 1987.
13. **Cavalli Sforza, L. L. and Edwards, A. W. F.,** Phylogenetic analysis: models and estimation procedures, *Evolution,* 21, 550, 1967.
14. **Malacrida, A. R., Guglielmino, C. R., Gasperi, G., Baruffi, L., Villani, P. C., and Milani, R.,** Genetical approach to systematics and phylogeny of Trypetinae (Diptera, Tephritidae), *Boll. Zool.,* 58, 355, 1991.
15. **Malacrida, A. R., Gasperi, G., and Milani, R.,** Genome organization of *Ceratitis capitata*: linkage groups and evidence for sex-ratio distorters, in *Fruit Flies,* Economopoulos, A. P., Ed., Elsevier Science Publ., Amsterdam, 1987, 169.
16. **Hancock, D.,** Ceratitinae (Diptera, Tephritidae) from Malagasy subregion, *J. Entomol. Soc. Sth. Afr.,* 47, 277, 1984.

Chapter 3

FOOD FORAGING BEHAVIOR OF FRUGIVOROUS FRUIT FLIES

J. Hendrichs and R. J. Prokopy

I. RESOURCE FORAGING BEHAVIOR

To manage agricultural pests more effectively it is useful to have a thorough understanding of their foraging behaviors for such vital resources as food, mates, oviposition sites, and shelter. A fundamental question of how organisms adjust their activities in response to the changing nature and distribution of resources in behavioral ecology is addressed by resource foraging theory.[1-4] One basic assumption underlying the study of resource foraging is that changes in the behavior of an organism as a function of the nature and spatial-temporal distribution of potential resources affect the organism's foraging efficiency and ultimately its fitness.

Providing specific or even hypothetical answers to some of the main questions addressed by studies of foraging behavior (Table 1) has in general proven challenging.[5] Apart from vertebrates, the considerable advances made to date in the study of foraging behavior have generally been restricted to model insects: for example, dung flies have been the object of mate foraging, bees of food foraging, and parasitoids of egg-laying site foraging studies.[4] For a majority of agricultural pests, however, progress in the quantitative analysis of foraging behavior has been limited. The reason is a lack of background information (of the type listed in Table 2), on the ecology, behavior, and physiology of the selected target organism under study.[5] Information of this type is indispensable for analyzing, in a meaningful way, the foraging behavior of a specific agricultural pest.

II. RESOURCE FORAGING INFORMATION

There is a considerable wealth of anecdotal and descriptive knowledge on adult behaviors for some tephritid species of economic importance. Quantitative foraging studies[5,6] have been restricted mostly to gaining information at the population level by recording fly distribution and movement patterns in space and time. Foraging studies quantifying behaviors at the level of individual flies are much more limited, although considerable advances have been made on oviposition-site foraging behavior,[5] and to a lesser degree on mate foraging behavior.[7-9]

Food foraging is the least systematically studied adult behavior in tephritid flies.[5,6] As pointed out by Tsiropoulos,[10] the spatial and temporal

Table 1

Questions Addressed by Studies of Foraging Behavior

-    does a forager aim at maximizing time and energy to acquire resources, at minimizing variation in amount or quality of resources, or at minimizing exposure to biotic and abiotic risks?

-    how does a forager resolve trade-offs in satisfying all types of resource requirements?

-    how does a forager sample resources to reach decisions?

Table 2

Background Information Required to Analyze Foraging Behavior

-    activity pattern in space and time in nature

-    precise identification of each resource type used in nature

-    estimation of quantity, quality, and distribution of each resource

-    characterization of resource finding behavior

-    role of physiological state, genetic background, and learning capability

-    biotic and abiotic risk-factors that affect foraging behavior

activity patterns of feeding by tephritid flies in nature are still largely unknown, as are all the types of food resources utilized, the quantity, quality, and changing distribution of each resource in the original or commercial habitat, the effects of particular foods on fly fecundity and longevity, the specific food stimuli to which hungry flies respond, as well as natural enemies, competitors, and other factors that might influence food foraging behavior. This is surprising, considering that detection, monitoring, and control of fruit fly pests is still dependent, to an important degree, on food-baits and the application of bait-insecticide sprays, sometimes imposed against stiff environmental opposition. In this chapter, we summarize recent advances in assessing various aspects of food foraging behavior in frugivorous tephritids, placing them into the context of management of fruit fly pests.

III. SURVIVAL REQUIREMENTS FOR ADULT FLIES

Tephritid flies engage in diurnal foraging bouts daily and throughout adult life to find water and a source of energy. Both of these feeding activities are indispensable for survival.[11,12] Indeed, in species studied so far, a majority of flies, held without water and carbohydrate, dies after about 48 hours.[12-17] In cases where flies find only carbohydrate but no water, survival lasts an average of 12-24 hours longer (total of 2.5 to 3 days), presumably because flies utilize the metabolic water produced during carbohydrate breakdown.[12] When flies find only water, but no carbohydrate, the deprivation period before death is slightly longer, ca. between 3 and 4 days. Accordingly, finding water appears to be of more immediate urgency for survival than the source of energy; although Kaiser and Schneider[16] showed that this is not the case under high humidity conditions. Moreover, if a fly obtains water, it is able to engage in additional foraging by flight. Considerable desiccation occurs during flight because thoracic spiracles have to remain open.[18] Ripley et al.[19] found that water-deprived Mediterranean fruit flies, *Ceratitis capitata* (Wied.), initiate flights with high frequency in nature, particularly after dewless nights, taking off with warm winds from tree tops and drifting down-wind to find water and food. Under dry conditions, Cunningham et al.[20] have shown that baits in water traps are 20 times more attractive to various fruit fly species, including medflies, than in wet climates.

The time available to find water and carbohydrate is especially limited for flies shortly after emerging from the pupal stage and therefore represents a particularly critical period in an adult's life. Flies normally emerge during the first 3 to 4 hours of the light cycle.[21] They crawl up vegetation and often remain motionless for hours while they harden their wings and cuticle. Because flies normally do not forage for food at night, they often have only the daylight hours of the day following emergence available to find these resources.

IV. ANAUTOGENOUS AND AUTOGENOUS LIFE STRATEGIES

Unlike many other types of Diptera, tephritid females typically produce oocytes in successive stages of maturation and, therefore, lay a few eggs every day.[22] For this type of oogenesis (eggs are produced in about 4-6 days), a regular supply of amino acids, sterols, vitamins, and minerals[10,12] is often required, generally also on a daily basis.[23]

Whereas water and carbohydrate (non-reproductive diet) are general requirements for maintenance and longevity of adult tephritids, in autogenous species, amino acids, sterols, vitamins, and minerals are not required for ovarian development and oogenesis. In anautogenous species, ovarian development ceases at the so called "resting stage" if females are not fed a diet conducive to reproductive development. Growth of previtellogenic oocytes resumes upon ingestion of a protein meal. This important difference in nutritional requirements between autogenous and anautogenous tephritid females appears to be related to the life strategy followed by the particular tephritid species.

Table 3
Nutritional Requirements in Adult Tephritids
in Relation to Larval Resource Exploitation Strategies

	PULPY FRUITS - relatively poor substrates - short lived resources - rapid larval development due to predation by vertebrate seed dispersers	OTHER STRUCTURES - often rich substrates - relatively long lived resources - longer protection from predation - interactive fly-plant system
SPECIALISTS - Stenophagous or monophagous - uni - or bivoltine - temperate - resource-based mating system	**mostly anautogenous** (*Rhagoletis spp.*)	**often autogenous** (particularly seed feeders and ball makers) (*Tephritini spp.*)
GENERALISTS - polyphagous - multivoltine - tropical or subtropical - non resource-based mating system (leks)	**mostly anautogenous** (*Anastrepha spp.*) (*Bactrocera spp.*) (*Ceratitis spp.*)	

There are three basic types of life strategies used by tephritids to exploit resources, according to Zwoelfer (Table 3).[24] Flies utilizing the first strategy are opportunistic polyphagous exploiters of pulpy fruits. These generalist frugivores are multivoltine and have a high reproductive potential. Most species live in tropical and subtropical regions of the world, where successive larval host resources are used.[25] To allow bridging between host fruiting seasons, adults are relatively long lived and usually highly mobile. Both of these features, together with high fecundity, increase the importance of regular intake in the adult stage of amino acids and other nutrients.[24]

Fruit flies utilizing the second life strategy found in tephritids are also frugivorous, although specialized (i.e. stenophagous or monophagous) exploiters of pulpy fruits. Members of this group are mostly univoltine or bivoltine, and occur in temperate climates where they spend most of their life in pupal diapause in the soil.[26] Precision in synchronization of adult emergence with host availability is more important than high reproductive potential, longevity, and mobility. However, these specialized flies are also anautogenous, requiring regular intake of amino acids and other nutrients to attain sexual maturity and development of eggs. This is due to a lack of nutritional reserves carried over from the larval stage. Pulpy fruits are generally short-lived. Therefore, larvae need to develop rapidly on a nutritionally poor diet. Furthermore, they must leave mature fruit as early as possible to escape predation by frugivorous birds and mammals.[27]

A third group of tephritids are specialized exploiters of plant structures such as galls or inflorescences. These species are primarily from temperate regions of the New and Old World where they are more numerous than frugivorous flies.[24] Unlike frugivorous tephritids that do not destroy seeds of host fruit, these tephritids can cause damage to the host plant itself. They have even been deployed as biological control agents of weeds.[28,29] Because larvae exploit plant structures that are relatively long-lived, are generally richer in nutrients than pulpy fruit, and sometimes offer protection from predators, adults are often autogenous, requiring only water and carbohydrate for survival (Table 3).[30] Some frugivorous tephritids are also protein autogenous. For example, the larvae of the papaya fruit fly, *Toxotrypana curvicauda*,[31] and *Dacus longistylus*[32] are protected in relatively long lived fruits and have access to nutritionally rich seeds. Females of these two species have extremely long ovipositors that penetrate through the latex of green immature fruit. Eggs are laid into the protected hollow fruit center where initially larvae feed exclusively on the maturing seeds and only feed on the fruit flesh when the fruit has ripened and the latex has disappeared.

Interestingly, some species of anautogenous frugivorous tephritid pests have the potential for at least partial autogeny.[33,34] Typically, frugivorous flies go through a sexual maturation period, which is required for development of the reproductive organs and initiation of oogenesis.[22] Proteinaceous food is required at this time. Length of premating periods depend on the tephritid species and the larval host.[35] They last from 7 to 14 days or even longer if proteinaceous food is not readily found. On the other hand, the Mediterranean fruit fly, held under mass rearing conditions for many generations on a nutritionally rich larval diet, has a premating period of about to 1-3 days as opposed to 7-10 days in wild flies. Whereas wild medfly females cannot produce eggs when restricted to sucrose and water, most laboratory adapted females can maintain egg production for 1-2 weeks while feeding only on water and sucrose.

V. PATTERNS AND SITES OF FLY FEEDING ACTIVITIES IN NATURE

Numerous workers have made observations on food sources encountered and consumed by adult tephritids in nature.[36] In addition, in various field studies aimed at understanding the behavior of fly activities in nature, some observations on feeding times and sites have been recorded in *Rhagoletis*,[37-40] *Anastrepha*,[41,42] and *Bactrocera (Dacus)* species.[36]

Food foraging behavior was quantified systematically in three studies of anautogenous frugivores by recording locations and diel patterns of sites and times of adult feeding from sunrise to sunset during a fly generation. The first study with the medfly was carried out under high population density conditions in a semi-isolated mixed orchard and surroundings in southern Egypt.[43] The second study was conducted under low population density conditions, and took place in an orange grove and surroundings on the island of Chios in Greece.[44] The third feeding study, with the apple maggot fly *Rhagoletis pomonella* (Walsh), was undertaken in an abandoned apple orchard and surroundings in eastern North America.[45] In all of these field studies, observations occurred on fruiting host trees, non-fruiting host trees, and surrounding non-host vegetation. For both species, males and females were found at dawn to be resting in upper sunlit parts of host and non-host tree canopies. Here, flies initiated feeding on droplets of dew. They moved progressively to lower, more shaded areas of the canopy with increasing temperature and light. Females of both species dispersed and fed more often than males. Females invested considerable time and energy foraging for food throughout most of the day, often away from fruiting host plants. They scavenged for nitrogenous food sources (fresh and dry bird droppings) both on host and non-host plant foliage (Table 4). The use of bird feces for food is not surprising considering that nitrogen is such a limiting resource for most phytophagous insects. Adults of numerous insects also feed on them extensively.[46-48]

Honeydew has been widely considered the normal source of most proteinaceous nutrients of adult tephritid flies in nature.[18,49,50] Fly populations observed in all field studies were largely sustained by nutrients other than insect honeydew. Honeydews are relatively low in nitrogen and lack some essential amino acids,[51,52] therefore, many tephritid adults obtain and possibly even require substantial nutrients from other sources of natural food to realize peak fecundity under natural conditions.

One of the main differences in food foraging found between the medfly and the apple maggot fly in these studies was the site of feeding. Wounded fruit and juices oozing from ripe fruit were the most common feeding sites for medflies and are probably sources of water and nutrients. Feeding on fruit was relatively uncommon in apple maggot flies, which fed mostly on leachates on leaf surfaces of host and non-host vegetation and only occasionally on fruits and nectars of flowers (Table 4).

Differences between medflies and apple maggot flies appear linked to the distribution of foraging for other resources and the type of mating system which is resource-based in the apple maggot fly and non-resource-based lek

Table 4

Plant Structures and Substrates on Fruiting and Non-fruiting hosts
and Non-hosts on which A) Flies of a Temperate Specialist Frugivore
(*R. pomonella)* and B) Flies of a Tropical Generalist Frugivore
(C. capitata) were Observed Feeding Over One Fly Generation

Percentage of Fly Feeding

	FOLIAGE	FRUIT	OTHER	TOTAL
A) *R. Pomonella*[45]				
Undefined matter on plant surfaces	68	9	2	79
Bird droppings	10	1	1	12
Fruit juice	2	5	1	8
Other	0	1	0	1
Total (n=410)	80	16	4	100
B) *C. Capitata*[43]				
Undefined matter on plant surfaces	20	6	1	27
Bird droppings	10	0	1	11
Fruit juice	0	60	0	60
Other	0	2	0	2
Total (n=423)	30	68	2	100

mating system in the medfly. Both sexes of medfly shift daily between
lekking, oviposition, and feeding sites on hosts and non-hosts. Similar daily
shifts in movement between commercial hosts and nearby wild vegetation and
between types of activity have been reported in other tropical frugivores.[36,42]

VI. ASSESSMENT OF NUTRITIVE VALUE OF NATURAL FOODS

Studies directly assessing natural foods for their contribution to fly
survival and egg development have been few and generally limited.[15,50,53] An
exception is a study by Tsiropoulos[54] with the olive fly *Dacus oleae*, although
this study did not involve field observations.

In the three field studies referred to above, the principal natural foods
identified were assessed for their contribution to egg-laying and energetic
maintenance. For medflies, results indicate that fruit such as grapes did not
support egg development, contributing only to longevity.[44] Fig fruit, with a
higher content of protein than most pulpy fruits, sustained both longevity and
fecundity. Bird feces alone supported neither egg production nor longevity.

However, when added to a diet of figs, bird feces significantly increased fly fecundity.

Results of a series of field cage tests for apple maggot flies with potted host trees, indicated that fly survival can be sustained by carbohydrates obtained from host foliage surfaces apparently in the form of plant leachate.[55] This finding perhaps helps to explain the often observed extensive "grazing" by apple maggot flies (in the absence of insect honeydew) on non-visible substances on host plant surfaces. Adult fecundity was not sustained by host foliage leachate.   Preparations of leaf surface bacteria (*Bacillus* sp., *Enterobacter* sp., *Micrococcus* sp.), pollen, insect frass, and uric acid (the main component of bird droppings), did not support significant egg-development, whereas bird droppings (most likely fecal bacteria), aphid honeydew, and to a considerably lesser extent, hawthorn fruit, did sustain egg development, though at a level significantly below that of enzymatic yeast hydrolysate.

Flies feed in alternating fashion (perhaps based on physiological need) on a variety of substrates in nature, some of which provide mostly water and carbohydrate for survival and others which provide mostly nutrients for egg development. Future studies to assess the nutritive contribution of natural foods to fly survival and fecundity should therefore allow for diet balancing (i.e. self-selection by flies, preferably in large field cages with trees, from a combination of identified natural food sources). Flies may achieve an appropriate nutrient mix by foraging for and responding to specific food odors which they associate, either instinctively or through learning, with the presence of a required nutrient.[56] By responding to ammonia and a number of other volatiles, usually hydrolytic, oxidative or microbial breakdown products of natural foods,[57-66] females may reduce time and energy spent foraging for dispersed sources of food and increase time available to forage for fruit in which to oviposit.

VII. FORAGING AND FOOD HANDLING AS AFFECTED BY FOOD QUALITY AND QUANTITY

Detailed mechanistic studies of dipteran neuro-physiological responses and feeding behaviors are described in Dethier's[67] classic, the "Hungry Fly." Dipteran movement within and between food patches that vary in distribution, quality, and quantity has been addressed in laboratory studies by Bell[68,69] and references therein. We have only recently initiated foraging studies of individual tephritid flies while they forage for food resources. In one series of field cage tests, we assessed feeding, food handling, and post-feeding foraging activities of individual apple maggot flies.[70] The purpose was to establish (under constant levels of previous food deprivation) food acceptance and ingestion thresholds and to contribute to the understanding of the dynamics of fly intra-tree foraging behaviors as affected by foods of varying quality and quantity, as well as the physiological state of the fly. Feeding and post-feeding behaviors were recorded for each fly after it was presented with yeast hydrolysate or sucrose particles or droplets, varying either in concentration, amount of food solute or total droplet volume.

Fly foraging time ("giving up time") on a tree branchlet was positively related to total amount of food solute previously encountered on a leaf surface and was largely independent of food volume or concentration for both carbohydrate and protein substrates (Table 5a). The volume and concentration of food presented, however, affected food "handling" and "processing" time significantly and, therefore, food foraging efficiency. In fact, total patch (branchlet) residence time was more closely linked to food handling and processing time than to food foraging time (Table 5b). Less time was needed for uptake of liquid than dry food. The latter required slow liquification by salivary secretion and elicited considerable intermittent cleaning of mouthparts by feeding flies. Uptake time in *R. pomonella* decreased with increasing dilution, although below a threshold of ca. 30 percent concentration of solute, rate of nutrient intake decreased rapidly.

With increasing dilution of food and increasing total food volume ingested, engorged flies entered extended quiescent post-feeding periods ("food processing time") during which they engaged in oral extrusion of droplets of liquid crop contents ("bubbling"). Afterwards, they reinitiated feeding, followed by more bubbling and feeding bouts. In a follow-up study,[71] we confirmed that bubbling is not a mechanism to achieve evaporative cooling, but rather a food "processing" behavior to eliminate excess water. Fly weight loss while bubbling was significantly correlated with duration of bubbling, temperature, and relative humidity during post-feeding and to initial fly weight (Adj.R^2 = 0.95) and was an order of magnitude higher than pre-feeding weight loss.

The methodology developed and knowledge gained in this type of intra-tree foraging study should facilitate future analyses of the dynamics of fly foraging behavior under interactive food, mate, and oviposition site resource conditions.

VIII. INTRA-TREE FORAGING FOR COMMERCIAL BAITS IN COMPETITION WITH NATURAL FOODS

Foraging studies are particularly important from an applied point of view where artificial foods, such as used in bait-insecticide sprays or food trap baits, are in competition with the natural foods. The apparent importance of bird droppings as natural fly food in our field observations,

Table 5a
Effect of a Constant Quantity (Solute Weight) of Yeast Hydrolysate Presented in Droplet Sizes of Increasing Volume and Decreasing Concentration on Foraging of Individual *R. pomonella* Flies[a]

| Droplet Volume (ul) | Solute | | Flies Ingesting Entire Droplet (%) | Time (min) | | Leaves Visited (Number) |
	Conc. (%)	Weight (ug)		Feeding	Foraging	
0.15	100.00	125.0	97 a	2.70 a	4.7 a	13.3 a
0.50	30.00	125.0	100 a	0.64 bc	5.2 a	11.8 a
1.00	15.00	125.0	97 a	0.88 bc	5.1 a	9.7 ab
2.00	7.50	125.0	93 a	1.51 b	4.3 a	6.0 bc
4.00	3.75	125.0	83 ab	2.68 a	4.6 a	4.4 bc
8.00	1.87	125.0	67 b	2.86 a	3.6 ab	3.2 c
0.00	0.00	0.0	0 c	0.00 c	1.8 b	4.2 bc

[a]   Comparison of means by Tukey's HSD-test ($p < 0.05$). Thirty flies tested for each treatment.[70]

Table 5b
Effect of a Constant Quantity (Solute Weight) of Yeast Hydrolysate Presented in Droplet Sizes of Increasing Volume and Decreasing Concentration on Food Ingestion, Subsequent Resting, and Regurgitation Behavior (Bubbling) of *R. pomonella* Flies[a]

| Droplet Volume (ul) | Solute | | Flies Ingesting Entire Droplet (%) | Time (min) | | Flies Bubbling % |
	Conc. (%)	Weight (ug)		Resting	Total	
0.15	100.00	125.0	97 a	2.8 c	10.3 c	3 a
0.50	30.00	125.0	100 a	4.8 c	10.6 c	20 a
1.00	15.00	125.0	97 a	13.9 b	20.0 b	83 b
2.00	7.50	125.0	93 a	20.6 a	26.4 a	100 b
4.00	3.75	125.0	83 ab	19.7 a	27.0 a	97 b
8.00	1.87	125.0	67 b	19.5 a	26.0 a	87 b
0.00	0.00	0.0	0 c	1.4 c	3.2 d	0 a

[a]   Comparison of means by Tukey's HSD-test ($p < 0.05$). Thirty flies tested for each treatment.[70]

Table 6
Responses of Individual Wild *C. capitata* Flies of Different Age and Days of Protein Deprivation During 15 Min Observation Periods When Released Onto the Canopy of Field-caged Guava Trees[a]

	Percentage Released *C. Capitata* Flies Arriving		
Age (Days) at Testing and Days of Protein Deprivation	Bird Feces	80% PIB-7 20% Water	Water
2 (n=48)	18 b 1	2 a 2	0 a 2
7 (n=48)	31 ab 1	4 a 2	2 a 2
12 (n=48)	42 a 1	6 a 2	2 a 2

[a] Values in each column (row) followed by the same letter (number) are not significantly different at the 0.05 level (G-test with Yates correction).[72]

Table 7
Responses of Individual Protein-deprived 5-day-old Wild *C. Capitata* Flies During 15 Min Observation Periods, When Released Onto the Canopy of a Field-caged Guava Tree[72]

	Percentage of Released *C. capitata* Flies		
	Arriving at Single Wick of Bird Feces	Arriving at 20 3-mm Droplets of 80% BIP-7 Bait and 20% Malathion ULF	Leaving Tree During Observation Period
No Choice Test (n=56)	-	4	61
Choice Test (n=56)	54	0	46

and the relationship of seed-dispersing frugivorous birds and mammals with fruiting plants appears to play a fundamental role in the ecology and evolution of frugivorous insects.[73] Therefore, we initiated food foraging studies comparing bird feces as natural food with commercial bait. In field-cage experiments with host trees, the proportion of medflies arriving at bird feces within 15 minute test periods was significantly greater (at least 7 times greater) than the proportion arriving at the commercial PIB-7 bait or water.[72] This outcome held irrespective of fly age (Table 6). In further tests that mimicked the size, density, and distribution of PIB-7 bait spray droplets on tree foliage, typical of an aerial medfly control program, very few (4%) or no released protein-deprived wild medflies found a bait droplet even though a majority (54%) found a single deposit of bird feces (Table 7).

These results show that aerial bait sprays against medflies might be improved substantially and the proportion of area treated with bait spray reduced considerably by (1) including synthetic equivalents of attractive components of bird feces in the spray mixture, and (2) adjusting spatial and temporal patterns of bait spray applications according to estimates of the composition and abundance of natural medfly food in infested regions. In another field-cage study with medflies,[74] droppings from domestic and feral birds and lizards were shown to be significantly more attractive to medflies than droppings from domestic mammals. Digestive systems, diets, and associated bacteria may contribute substantially to differences in attractiveness among droppings.

Similar results were found under natural, semi-natural, and field conditions with the apple maggot fly.[75] Droppings from birds fed antibiotics or droppings treated with antibiotics were less attractive than droppings not affected by antibiotics, thereby confirming that bacteria are involved in generating the attractive volatiles. It is known that both anaerobic bacteria[76] and facultative aerobic bacteria[48] are capable of decomposing uric acid in bird droppings. Perhaps it is because of multiplication of bacteria with time that volatiles from slightly aged droppings from birds receiving high protein diets, appear to be the most promising for developing improved lures.

IX. FOOD FORAGING IN RELATION TO PEST MANAGEMENT

Adults of autogenous species, such as the papaya fruit fly, do not represent the typical "protein-hunger" pattern of maturing and mature anautogenous flies.[31,77] They do not respond to proteinaceous food lures or baits as do anautogenous flies. Fortunately, a majority of frugivorous fruit fly pests of economic importance are anautogenous. They respond to a number of commercial food attractants that have been in use since the beginning of the century as food baits for monitoring and control of these pests.[78]

Presently, most effective non-food attractants in use for the management of tephritid pests are male-specific, whereas the attraction of females is still largely dependent on food baits.[79] Efforts to develop effective host or sexual female attractants have, with some important exceptions,[80,81] not been successful. As shown by the foraging studies above, bird droppings with fecal bacteria competed successfully for foraging females with widely used commercial baits.[72,75] The possibilities for more selective, potent or practical food attractants for females are far from exhausted. Important priorities in the development of improved female attractants are efforts to refine present commercial food baits,[58,82-84] to identify active components and to prepare formulations from natural sources of food volatiles such as bird

Table 8
Some Orchard Management and Fruit Fly Control Measures Under Consideration that Relate to Adult Food Foraging Behavior

1. **Reduction of natural food sources in orchards by:**
    - adjusting pruning practices to remove excess fresh succulent foliar growth (e.g. watersprout) and attendant aphids
    - discouraging flocks of birds from entering orchards through use of Scare-eye balloons
    - harvesting fruit slightly before maturity and removing wounded or fallen fruit

2. **Control of fruit flies in such food-scare orchards by:**
    - placing food/water baited traps around orchards to intercept incoming flies
    - placing such traps in the perimeter rows of trees to capitalize on the need of flies to move regularly to surrounding vegetation to obtain food
    - confining bait-sprays, when required, to orchard perimeters where fallen and fermenting fruit is not removed and is used as a good trap

droppings. Volatiles from other natural products such as fermenting host fruit,[64] bacteria on plant surface substrates,[59,63,65,66] vertebrate urine,[61] and various types of insect honeydews are also excellent candidates for study.

Some of the recent foraging studies reviewed have potentially strong practical impact on strategies and tactics for managing frugivorous flies. Under mixed host conditions in a natural setting, flies adjust their food foraging activities in response to dynamic changes in the spatial, temporal, and seasonal distribution of food resources and host phenologies. Under these conditions, the effectiveness of food baits is quite variable due to the influence of the type and amount of natural food present. Food foraging is probably less complex and more predictable under commercial orchard monoculture conditions. Here, a number of orchard management practices that involve fly food foraging behavior can be considered (Table 8). Harvesting fruit slightly before maturity, removing wounded or fallen ripe fruit, or adjusting pruning practices to discourage formation of fresh watersprouts and attendant buildup of aphids are possibilities for maintaining plantations or orchards comparatively free of some important food resources. Furthermore, discouraging flocks of birds from entering orchards through the use of Scare-eye balloons would result in fewer wounded fruit and fewer bird droppings as sources of food for adults. In such food-scarce commercial plantations or orchards, one might expect that immigrating flies remain primarily in the perimeter rows of trees because of their need to move back and forth regularly to the surrounding vegetation to obtain food. Such an obligatory movement would increase the effectiveness of food-baited traps placed around orchards. Such traps would be particularly valuable in intercepting immature females before they cause damage at a time when their response to host- and pheromone-type lures is still weak. Possibly as important, the widely used ground or aerial insecticide bait sprays would become more effective in the face of reduced competition from natural food.

Bait-sprays could be confined specifically to the perimeters of food-scarce orchards, thereby contributing successfully to a more environmentally oriented management of these pests. Some of these possibilities are already being tested in Massachusetts apple orchards[85-87] to assess the effects on fruit fly control. Examination of food foraging behavior and testing of these possibilities for fly management in other important agroecosystems favored by flies (i.e. coffee plantations or citrus orchards) should be the subject of future studies.

Food foraging behaviors may also play an important role in relation to the implementation of the sterile insect technique (SIT). Pre-SIT application of insecticidal food baits is recommended for suppressing large fly populations to achieve adequate overflooding ratios. For low pest population densities, Knipling[88] has shown that one bait spray before release of sterile insects increases dramatically the impact of sterile flies by eliminating mated females that would otherwise continue ovipositing fertile eggs for one generation.

Implications of food foraging and feeding by sterile flies for SIT programs is another area that deserves close attention. An adequate level of sexual performance in laboratory reared strains of flies cannot be exploited to the full extent if released individuals fail to disperse and perform properly in the field because they are unable to locate food sources.[89] Although sterile flies are generally considered to require less food than non-irradiated wild flies,[16,90] it was recently shown by Landolt and Sivinski[91] that sterile *Anastrepha suspensa* flies, when deprived of sugar during limited hours preceding the sexual activity period, were unable to maintain pheromone calling during that period. As a result, most sexual activity of sterile males ceased. However, when these sterile males found overripe host fruit on which to feed, their capacity of pheromone calling was reestablished. They concluded that the distribution of natural sources of carbohydrate in nature may profoundly affect survival and distribution of *A. suspensa* males, influence the selection of male calling sites, and affect male reproductive success. Plant and Cunningham[92] reported that high mortality of released sterile medflies is associated with periods when orchard trees are not flowering. They imply that flowers provide a vital source of food for the released population in the absence of other natural food sources under Hawaiian conditions. When food is scarce in nature, consideration could be given to supplement sterile fly releases with sources of carbohydrate, as recommended in some augmentative biological control programs.[93] Monro[94,95] proposed an opposite approach, suppressing or eradicating fly populations by overloading limited food resources available to wild flies by the massive release of sterile flies.

X. CONCLUSION

In this review, we have discussed various aspects of the food foraging behavior of tephritid flies that are relevant to life history strategies, trophic relations, and approaches to pest management. We conclude that present knowledge of food foraging behavior in nature and the contribution of various natural foods to survival and reproduction is relatively weak. Future study of tephritid food foraging behavior could have a strong impact on adjusting existing tactics and developing new approaches to fly management.

REFERENCES

1. Hassel, M. P. and Southwood, T. R. E., Foraging strategies of insects, *Ann. Rev. Ecol. Syst.*, 9, 75, 1978.
2. Kamil, A. C. and Sargent, T. D., *Foraging Behavior: Ecological, Ethological and Psychological Approaches*, Garland Press, 1981.
3. Pyke, G., Optimal foraging theory: a critical review, *Ann. Rev. Ecol Syst.*, 15, 532, 1984.
4. Stephens D. W. and Krebs, J. R., *Foraging Theory*, Princeton University Press, Princeton, New Jersey, 1986, pp. 247.
5. Prokopy, R. J. and Roitberg, B. D., Fruit fly foraging behavior, in *Fruit Flies, their Biology, Natural Enemies and Control, Vol. 3A*, Robinson, A. S. and Hooper, G., Eds., Elsevier Science Publishers, Amsterdam, 1989, 293.
6. Prokopy, R. J. and Roitberg, B. D., Foraging behavior of true fruit flies, *Amer. Scient.*, 72, 41, 1984.
7. Hendrichs, J., Sexual selection in wild and sterile Caribbean fruit flies, *Anastrepha suspensa* (Diptera: Tephritidae), M.Sc. Thesis, University of Florida, Gainesville, 1986, pp. 263.
8. Opp, S. B. and Prokopy, R. J., Seasonal changes in resightings of marked, wild *Rhagoletis pomonella* (Diptera: Tephritidae) flies in nature, *Fla. Entomol.*, 70, 449, 1987.
9. Robacker, D. C., Mangan, R. L., Moreno, D. S., and Tarshis Moreno, A. M., Mating behavior and male mating success in wild *Anastrepha ludens* (Diptera: Tephritidae) on a field-caged host tree, *J. Insect Behavior*, 4, 471, 1991a.
10. Tsiropoulos, G. J., The role of feeding behavior in fruit fly population management, in *Fruit Flies of Economic Importance* 87, Cavalloro, R., Ed., A. A. Balkema, Rotterdam, 1989, 129.
11. Webster, R. P., Stoffolano, Jr., J. G., and Prokopy, R. J., Long-term intake of protein and sucrose in relation to reproductive behavior of wild and laboratory cultured *Rhagoletis pomonella*, *Ann. Entomol. Soc. Amer.*, 72, 41, 1979.
12. Tsitsipis, J. A., Nutrition Requirements, in *Fruit Flies, their Biology, Natural Enemies and Control, Vol. 3B*, Robinson, A. S. and Hooper, G., Eds., Elsevier Science Publishers, Amsterdam, 1989, 103.
13. Mason, A. C., Biology of the papaya fruit fly, *Toxotrypana curvicauda* in Florida, *U.S. Dept. Agric. Bull.*, 1081, 1, 1922.
14. Middlekauff, W. W., Some biological observations of the adults of the apple maggot and cherry fruit flies, *J. Econ. Entomol.*, 34, 621, 1941.

15.   **Marlowe R. H.**, Effect of foods on ovarian development in the melon fly, *J. Econ. Entomol.*, 38, 339, 1945.

16.   **Kaiser, I. and Schneider, E. L.**, Need for immediate sugar and ability to withstand thirst by newly emerged oriental fruit flies, melon flies, and Mediterranean fruit flies untreated or sexually sterilized with gamma radiation, *J. Econ. Entomol.*, 62, 539, 1969.

17.   **Hendrichs, J. and Prokopy, R. J.**, unpublished data.

18.   **Downes, Jr., W. L. and Dahlem, G. A.**, Keys to the evolution of Diptera: role of Homoptera, *Environ. Entomol.*, 16, 847, 1987.

19.   **Ripley, L. B., Hepburn, G. A., and Andersen, E. E.**, Fruit fly migration in the Kat River valley, *South African Dept. Agric. and Forestry, Plant Industry Series*, No. 49, 4, 1940.

20.   **Cunningham, R. T., Nakagawa, S., Suda, D. Y., and Urago, T.**, Tephritid fruit fly trapping: liquid food baits in high and low rainfall climates, *J. Econ. Entomol.*, 71, 762, 1978.

21.   **Smith, P. H.**, Behavioural partitioning of the day and circadian rhythmicity, in *Fruit Flies, their Biology, Natural Enemies and Control, Vol. 3A*, Robinson, A. S. and Hooper, G., Eds., Elsevier Science Publishers, Amsterdam, 1989, 325.

22.   **Williamson, D. L.**, Oogenesis and spermatogenesis, in *Fruit Flies, their Biology, Natural Enemies and Control, Vol. 3B*, Robinson, A. S. and Hooper, G., Eds., Elsevier Science Publishers, Amsterdam, 1989, 141.

23.   **Hendrichs, J., Cooley S. and Prokopy, R. J.**, How often do apple maggot flies need to eat?, *Mass. Fruit. Notes.*, 55(3), 12, 1990a.

24.   **Zwoelfer, H.**, Life systems and strategies of resource exploitation in tephritids, in *Fruit Flies of Economic Importance*, Cavalloro, R., Ed., A. A. Balkema, Rotterdam, 1983, 16.

25.   **Bateman, M. A.**, The ecology of fruit flies, *Ann. Rev. Entomol.*, 17, 493, 1972.

26.   **Boller, E. F. and Prokopy, R. J.**, Bionomics and management of *Rhagoletis*, *Ann. Rev. Entomol.*, 21, 223, 1976.

27.   **Drew, R. A. I**, Reduction in fruit fly (Tephritidae: Dacinae) populations in their endemic rainforest habitat by frugivorous vertebrates, *Aust. J. Zool.*, 35, 283, 1987

28.   **Zwoelfer, H.**, Evolutionary and ecological relationships of the insect fauna of thistles, *Ann. Rev. Entomol.*, 33, 103, 1988.

29.   **Harris, P.**, The use of Tephritidae for the biological control of weeds, *Biocontrol News and Information*, 10, 7, 1989.

30.   **Vogt, E. A.**, Influence of adult diet on oogenesis, longevity, and fecundity of *Urophora affinis* Fraunfeld, with notes on *U. quadrifasciata* Meigen (Diptera: Tephritidae), M.Sc. Thesis, University of Idaho, 1986, pp. 67.

31.   **Landolt, P. J.**, Reproductive maturation and premating period of the papaya fruit fly, *Toxotrypana curvicauda* Gerstaecker (Diptera: Tephritidae), *Fla. Entomol.*, 67, 240, 1984.

32.   **Hendrichs, J. and Reyes, J.**, Reproductive behavior and post-mating female guarding in the monophagous multivoltine *Dacus longistylus* (Diptera: Tephritidae) in southern Egypt, in *Fruit Flies*, Economopoulos, A. P., Ed., Elsevier, Amsterdam, 1987, 303.

33.   **Bustamante, R. E.**, Efecto del alimento en la supervivencia y fecundidad de los adultos de la mosca del mediterraneo *Ceratitis capitata* (Wied.)(Diptera: Tephritidae), B. S. Thesis, Escuela de Biologia, Universidad Autonomede Chiapas, Tuxtla Gutierrez, 1992.

34.   **Hendrichs, et al.**, unpublished data.

35.   **Krainacker, D. A., Carey, J. R., and Vargas, R. I.**, Effect of larval host on life history traits of the mediterranean fruit fly, *Ceratitis capitata*, *Oecologia*, 73, 583, 1987

36.   **Nishida, T.**, Food systems of tephritid fruit flies in Hawaii, *Proc. Entomol. Soc. Hawaii*, 23, 245, 1980.

37. **Prokopy, R. J., Bennett, E. W., and Bush, G. L.,** Mating behavior in *Rhagoletis pomonella.* II. Temporal organization, *Can. Entomol.,* 104,97, 1971.

38. **Prokopy, R. J.,** Feeding, mating and oviposition activities of *Rhagoletis fausta* in nature, *Ann. Entomol. Soc. Amer.,* 69, 899, 1976.

39. **Smith, D. C. and Prokopy. R. J.,** Seasonal and diurnal activity of *Rhagoletis mendax* flies in nature, *Ann. Entomol. Soc. Amer.,* 74, 462, 1981.

40. **Frias, D.,** Field observations of distribution and activities of *Rhagoletis conversa* on two hosts in nature, *Ann. Entomol. Soc. Amer.,* 77, 548, 1984.

41. **Burk, T.,** Behavioral ecology of mating in the Caribbean fruit fly, *Anastrepha suspensa* (Loew) (Diptera: Tephritidae), *Fla. Entomol.,* 66, 330, 1983.

42. **Malavasi, A., Morgante, J. S., and Prokopy, R. J.,** Distribution and activities of *Anastrepha fraterculus* (Diptera: Tephritidae) flies on host and non-host trees, *Ann. Entomol. Soc. Amer.,* 76, 286, 1983.

43. **Hendrichs, J. and Hendrichs, M. A.,** Mediterranean fruit fly (Diptera: Tephritidae) in nature: location and diel pattern of feeding and other activities on fruiting and non-fruiting hosts and non-hosts, *Ann. Entomol. Soc. Amer.,* 83, 632, 1990.

44. **Hendrichs, J., Katsoyannos, B. I., Papaj, D. R., and Prokopy, R. J.,** Sex differences in movement between natural feeding and mating sites and tradeoffs between food consumption, mating success and predator evasion in Mediterranean fruit flies (Diptera: Tephritidae), *Oecologia,* 86, 223, 1991.

45. **Hendrichs, J. and Prokopy, R. J.,** Where do apple maggot flies find food in nature?, *Mass. Fruit. Notes,* 55(3), 1, 1990.

46. **Adler, P. H. and Wheeler, Jr., A. G.,** Extra-phytophagous food sources of Hemiptera-Heteroptera: bird droppings, dung and carrion, *J. Kansas Entomol. Soc.* 57, 21, 1984.

47. **Ray, T. S. and Andrews, C. C.,** Antbutterflies: butterflies that follow army ants to feed on antbird droppings, *Science,* 210, 1147, 1980.

48. **Schal, C. and Bell, W. J.,** Ecological correlates of paternal investment of urates in a tropical cockroach, *Science,* 218, 170, 1982.

49. **Hagen, K. S.,** Honeydew as an adult fruit fly diet affecting reproduction, *Proc. Tenth Int. Congress of Entomology,* Montreal 1956, 3, 25, 1958.

50. **Neilson, W. T. A. and Wood, F. A.,** Natural source of food of the apple maggot, *J. Econ. Entomol.,* 59, 997, 1966.

51. **Craig, R.,** The physiology of excretion in the insect, *Ann. Rev. Entomol.,* 5, 53, 1960.

52. **Van Vianen, A.,** Honeydew - a historic overview, *Med. Fac. Landbouww. Rijsksuniv. Gent.,* 54, 955, 1989.

53. **Christenson, L. D. and Foote, R. H.,** Biology of fruit flies, *Ann. Review Entomol.,* 5, 171, 1960.

54. **Tsiropoulos, G. J.,** Reproduction and survival of the adult *Dacus oleae* (Gmel.) feeding on pollens and honeydews, *Environ. Entomol.,* 6, 390, 1977.

55. **Hendrichs, J., Lauzon, C. R., Cooley, S., and Prokopy, R. J.,** Contribution of natural food sources to the longevity and fecundity of *Rhagoletis pomonella* flies (Diptera: Tephritidae), *Ann. Entomol. Soc. Amer.,* 1993 (In Press).

56. **Robacker, D. C.,** Specific hunger in *Anastrepha ludens* (Diptera: Tephritidae): effects on attractiveness of proteinaceous and fruit-derived lures, *Environ. Entomol.,* 20, 1680, 1991.

57. **Gow, P. L.,** Proteinaceous bait for the Oriental fruit fly, *J. Econ. Entomol.,* 47, 153, 1954.

58. **Bateman, M. A. and Morton, T. C.,** The importance of ammonia in proteinaceous attractants for fruit flies (Family: Tephritidae), *Aust. J. Agr. Res.,* 32, 883, 1981.

59. **Drew, R. A. I. and Lloyd, A. C.,** Bacteria associated with fruit flies and their host plants, in *Fruit Flies, their Biology, Natural Enemies and Control, Vol. 3A,* Robinson, A. S. and Hooper, G., Eds., Elsevier Science Publishers, Amsterdam, 1989, 131.

60.   **Drew, R. A. I., Courtice, A. C., and Teakle, D. S.,** Bacteria as a natural source of food for adult fruit flies (Diptera: Tephritidae), *Oecologia,* 60, 279, 1983

61.   **Hedstroem, I.,** Una sustancia natural en la captura de moscas de la fruta del genero *Anastrepha* Schiner (Diptera: Tephritidae), *Rev. Biol. Trop.,* 36, 269, 1988.

62.   **Hendrichs, J., Hendrichs, M., Prokopy, J., and Prokopy, R. J.,** How do apple maggot flies detect the presence of distant food?, *Mass. Fruit Notes,* 55(3), 3, 1990b.

63.   **Jang, E. B. and Nishima, K. A.,** Identification and attractancy of bacteria associated with *Dacus dorsalis* (Diptera: Tephritidae), *Environ. Entomol.,* 19, 1726, 1990.

64.   **Robacker, D. C., Tarshis Moreno, A. M., Garcia, J. A. and Flath, R. A.,** A novel attractant for Mexican fruit fly, *Anastrepha ludens,* from fermented host fruit, *J. Chem. Ecol.,* 16, 2799, 1990.

65.   **Robacker, D. C., Garcia, J. A., Martinez, A. J., and Kaufman, M. G.,** Strain of *Staphylococcus* attractive to laboratory strain *Anastrepha ludens* (Diptera: Tephritidae), *Ann. Entomol. Soc. Amer.,* 84, 555, 1991b.

66.   **MacCollom, G. B., Lauzon, C. R., Weires, Jr., R. W., and Rutkowski, A. A.,** Attraction of adult apple maggot (Diptera: Tephritidae) to microbial isolates, *J. Econ. Entomol.,* 85, 83, 1992.

67.   **Dethier, V. G.,** *The Hungry Fly,* Harvard University Press, Cambridge, 1976.

68.   **Bell, W. J.,** Searching behavior patterns in insects, *Ann. Rev. Entomol.,* 35, 447, 1990.

69.   **Bell, W. J.,** *Searching Behavior: The Behavioral Ecology of Finding Resources,* Chapman and Hall, New York, 1991, pp. 345.

70.   **Hendrichs, J., Fletcher, B. S., and Prokopy, R. J.,** Post-feeding behavior of *Rhagoletis pomonella* flies: effect of initial food quantity and quality on food foraging, handling costs and bubbling, *J. Insect Behav.,* 1992a (In Press).

71.   **Hendrichs, J., Cooley, S. S., and Prokopy, R. J.,** Post-feeding bubbling behavior in fluid-feeding Diptera: concentration of crop contents by oral evaporation of excess water, *Physiol. Entomol.,* 17, 153, 1992b.

72.   **Prokopy, R. J., Papaj, D. R., Hendrichs, J., and Wong, T. T. Y.,** Behavioral responses of *Ceratitis capitata* flies to bait spray droplets and natural food, *Entomol. exp. and appl.,* 1992a (In Press).

73.   **Sallabanks R. and Courtney, S. P.,** Frugivory, seed predation and insect-vertebrate interactions, *Ann. Rev. Entomol.,* 37, 377, 1992.

74.   **Prokopy, R. J., Ling-Hsu, C., and Vargas, R. I.,** Effect of source and condition of animal excrement on attractiveness to *Ceratitis capitata* flies (Diptera: Tephritidae), *Environ. Entomol.,* 1993 (Submitted).

75.   **Prokopy, R. J., Cooley, S. S., Galarza, L, Bergweiler, C., and Lauzon, C. R.,** Bird droppings as natural food compete with bait sprays against *Rhagoletis pomonella* flies (Diptera: Tephritidae), *Canad. Entomol.,* 1992b (In Press).

76.   **Barnes, E. M. and Impey, C. S.,** The occurrence and properties of uric acid decomposing anaerobic bacteria in the avian caecum, *J. Appl. Bact.,* 37, 393, 1974.

77.   **Sharp J. L. and Landolt, P. J.,** Gustatory and olfactory behavior of the papaya fruit fly, *Toxotrypana curvicauda* Gerstaecker, (Diptera: Tephritidae) in the laboratory with notes on longevity, *J. Georgia Entomol. Soc.,* 19, 176, 1984.

78.   **Roessler, Y.,** Insecticidal bait and cover sprays, in *Fruit Flies, their Biology, Natural Enemies and Control, Vol. 3B,* Robinson, A. S. and Hooper, G., Eds., Elsevier Science Publishers, Amsterdam, 1989, 329.

79.   **Sivinski, J. M. and Calkins, C. O.,** Pheromones and parapheromones in the control of tephritids, *Fla. Entomol.,* 69, 157, 1986.

80.   **Fein, B. L., Reissig, W. H., and Roelofs, W. L.,** Identification of apple volatiles attractive to the apple maggot, Rhagoletis pomonella, *J. Chem Ecol.,* 8, 1473, 1982.

81.   Chuman, T., Landolt, P. J., Heath, R. R., and Tumlinson, J. H., Isolation, identification and synthesis of male-produced sex pheromone of papaya fruit fly, *Toxotrypana curvicauda* Gerstaecker (Diptera: Tephritidae), *J. Chem. Ecol.*, 13, 1979, 1986.

82.   Mazor, M., Gothilf, S., and Galun, R., The role of ammonia in the attraction of females of the Mediterranean fruit fly to protein hydrolyzate baits, *Entomol. exp. and appl.*, 43, 25, 1987.

83.   Wakabayashi, N. and Cunningham, R. T., Four-component synthetic food bait for attracting both sexes of the melon fly (Diptera: Tephritidae), *J. Econ. Entomol.*, 84, 1672, 1991.

84.   Teranishi, R., personal communication.

85.   Prokopy, R. J., Christie, M., Johnson, S. A., Rankin, K., and Donovan, C., Three years of the Massachusetts second stage apple IPM pilot project: blocks receiving apple maggot fly interception traps, *Mass. Fruit Notes*, 55(1), 4, 1990a.

86.   Prokopy, R. J., Johnson, S. A., and O'Brien, M. T., Second stage integrated management of apple arthropod pests, *Entomol. exp. and appl.*, 54, 9, 1990b.

87.   Duan, J. J. and Prokopy, R. J., Visual and odour stimuli influencing effectiveness of sticky spheres for trapping apple maggot flies *Rhagoletis pomonella* (Walsh) (Diptera: Tephritidae), *J. Appl. Entomol.*, 113, 271, 1992.

88.   Knipling, E. F., The Basic Principles of Insect Population Suppression and Management, Agriculture Handbook Number 512, United States Dept. Agric., Washington, D. C., 1979.

89.   Boller, E. F., Sexual activity, in *Quality Control, An Idea Book for Fruit Fly Workers*, Boller, E. F. and Chambers, D. L., Eds., West Paleartic Regional Section Bulletin, IOBC 1977/5, 1977, 75.

90.   Galun, R., Gothilf, S., Blondheim, S., and Lachman, A., Protein and sugar hunger in the Mediterranean fruit fly *Ceratitis capitata* (Wied.) (Diptera: Tephritidae), in *Determination of Behaviour by Chemical Stimuli*, Proc. 5th European Chemoreception Research Organization Symposium, 1981, 245.

91.   Landolt, P. J. and Sivinski, J., Effects of time of day, adult food, and host fruit on incidence of calling by male Caribbean fruit flies (Diptera: Tephritidae), *Environ. Entomol.*, 21, 382, 1992.

92.   Plant, R. E. and Cunningham, R. T., Analysis of the dispersal of sterile Mediterranean fruit flies (Diptera: Tephritidae) released from a point source, *Environ. Entomol.*, 20, 1493, 1991.

93.   Hagen, K. S., Sawall, Jr., E. F., and Tessan, R. L., The use of food sprays to increase the effectiveness of entomophagous insects, *Proc. Tall Timbers Conf. Ecol. Anim. Contr. by Habitat Management*, 2, 59, 1970.

94.   Monro, J., Population control in animals by overloading resources with sterile animals, *Science*, 140, 496, 1963.

95.   Monro, J., Population flushing with sexually sterile insects, *Science*, 151, 1536, 1966.

Chapter 4

ADVANCES IN ATTRACTANT AND TRAPPING TECHNOLOGIES FOR TEPHRITIDS

A. P. Economopoulos and G. E. Haniotakis

I. INTRODUCTION

Insect trapping is essential for population studies or for use in insect pest control programs. Estimation of population size, detection of newly introduced species and evaluation of population reproductive ability are necessary components for any control system. Attract and kill techniques are often used in insect pest management programs.

Attractants and traps have received much attention for fruit flies as components of fly management in small, or area-wide programs. Fruit flies are very mobile, have a high reproductive potential, and infest high value crops. Therefore, they are under strict quarantine regulations. Thus, intense suppression or eradication efforts are employed.

Some very powerful attractants have been discovered by chance, e.g. methyleugenol[1] and kerosene.[2] In more recent years, systematic studies of fruit fly behavior has led to substantial progress in developing trapping or control methods based on attraction.[3,4]

II. FOOD-ODOR ATTRACTANTS

A variety of food attractants, such as different protein hydrolysates and ammonium salts, have been used for trapping of both sexes of fruit flies. Ammonia appears to be the principal attractant originating from these food lures as found with *Dacus tryoni*.[5,6] Similar results were obtained with various protein hydrolysates tested with *Ceratitis capitata*, although there were indications that medfly response was affected by other volatiles as well.[7]

Amino acids in food attractants seem to elicit mainly phagostimulation[6] keeping the flies on the lure, thus causing death by increased consumption of insecticide or drowning in the case of McPhail traps. Ribonucleotides were also found to elicit phagostimulation in *Ceratitis capitata* and their possible synergistic effect with amino acids has been suggested.[8]

The odor(s) of fermented host fruit was found to elicit strong attraction in *Anastrepha ludens*. Three out of several chemicals identified were the main attractants acting additively, i.e. each elicited separate attraction. Their mixture was about twice as attractive as aqueous protein bait.[9] It has been suggested that odors emitted following injury to the fruit or fermentation could provoke *Ceratitis capitata* attacks.[10]

0-8493-4854-4/94/$0.00+$.50

Bird feces, a source of protein, was found to attract medflies strongly. Their attractive components could possibly increase bait spray effectiveness substantially if mixed with the bait.[11]

In recent years bacteria, originating from fruit fly alimentary canals, were found to elicit attraction of females, but at much lower levels than protein hydrolysates.[12-15]

Odors of protein-hydrolysates have been the most powerful attractants addressed to both sexes. Combination of food odor with other type attractants, e.g. color, parapheromones, and pheromones have often increased the number of flies that were attracted to traps.[16-19]

III. PHEROMONES

Considerable effort has been devoted to the search for pheromones, with sex attractants and aggregation pheromones receiving special attention. With only one known exception (the olive fruit fly), sex pheromones in fruit flies are produced by males and function as short to medium range female attractants. Field evaluations of these pheromones in most species, however, have revealed male attraction as well. Pheromone calling only occurs in mating arenas, referred to as leks. The males are attracted to the odor because this is where females come to be mated. The males either hope to intercept and mate with approaching females or to inject themselves into favorable positions in the lek. The following is a brief review of our recent knowledge for the most important fruit fly species.

A.   *CERATITIS CAPITATA*

Yang et al.[20] detected 69 compounds in volatiles collected from sexually mature, calling laboratory-reared male medflies. Of these, 56 were isolated and 54 were identified, 5 of which were major components. EAG studies of all 54 identified components as well as of 5 analogs, showed that significantly different responses between the sexes were found in 9 compounds, and that of the 5 major components, 3 elicited relatively small responses while 2 elicited large EAG responses. Behavioral bioassays of each of the 5 major components, as well as a blend of 6 of the compounds, showed some degree of attractiveness to virgin females, which in some cases approached the response to male odors absorbed into filter paper. Baker et al.[21] tested the synthetic components of those they consider the male medfly sex pheromones and reported that traps with linalool, 2,3-dimethylpyrazine, 2,5-dimethylpyrazine, or geranyl acetate were all more attractive to released sterile flies than unbaited traps, the sex ratios of catches varied greatly, and no component or mixture caught more flies than trimedlure. Heath et al.[22] identified three major components, ethyl-(E)-butanoate, geranyl acetate, and (E, E)-alfa-farnesene, from volatiles of male medflies. These three compounds were synthesized and formulated as a blend to be released in a ratio similar to that emitted by wild male flies. Attractiveness of the blend to female flies was demonstrated under field conditions. Nevertheless, it is clear that the desired pheromone for this species, a potent and selective attractant, if it exists, has not been found and the effort is continuing.

B. ANASTREPHA LUDENS

Battiste et al.[23] reported anastrephin and epianastrephin as novel lactone components isolated from the sex pheromone blend of the male Mexican and Caribbean fruit flies. Robacker and Heath[24] reported that (Z)-3 nonenol (a), (Z)-3,6 nonadienol (b), and (S, S)-(-)-epi-anastrephin (c) constitute the male produced pheromone of this species. All three compounds elicit attraction and/or locomotor arrest from virgin females when tested individually. When tested together (a) inhibits the effect of (b) but either of the alcohols synergizes the effects of (c). In field studies the pheromone attracts both sexes equally but yeast hydrolysate captures three times more flies than the most attractive pheromone dose (10 male equivalents).[25] Behavioral studies of virgin females to the pheromone showed that responses included attraction to the vicinity but not to the point sources, increasing searching rate, etc.[26] Combinations of pheromone with fermenting fruit odors gave results of questionable practical value.[27,28] Heath and Battiste[29] recently identified two new chemicals produced by calling males, thus raising the number of compounds produced by male Mexican fruit flies to seven.

C. ANASTREPHA SUSPENSA

The male pheromone of this species consists of four components, two 9-carbon alcohols containing one and two double bonds, respectively, and two lactone esters, anastrephin and epianastrephin.[30] The pheromone attracts virgin laboratory reared females of only a narrow age range of 9 to 12 days, a characteristic that will have severe practical limitations. Anastrephin and epianastrephin were isolated from both the Caribbean and Mexican fruit flies.[23] Nation,[31] in his studies on the biology of pheromone release by male Caribbean fruit flies, refers to seven compounds identified from the volatiles of males, two 9-carbon alcohols, three lactones, a sesquiterpene, and a monoterpene. As in the case of the previous fruit fly species, exposure of live males or of pheromones in the field, resulted in the capture of nearly equal numbers of males and virgin females. This is similar to the situation with medflies.

D. BACTROCERA (DACUS) OLEAE

Female olive fruit flies release a four-component pheromone blend[32] which functions as a long-range male attractant. The major component of the blend is 1,7-dioxaspiro[5.5]undecane, and a-pinene, nonanal and ethyl dodecanoate are the other three. It was found later that the R-(-)-enantiomer of the major pheromone is the long-range male attractant while the S-(+)-enantiomer is an aggregation pheromone for both sexes with additional arrestant and aphrodisiac properties.[33] Population monitoring[34,35] and control technologies have been developed that utilize pheromones. A mass trapping method, combining pheromones with a food attractant, a phagostimulant, and an hygroscopic substance, is currently being introduced gradually for commercial use[36] and constitutes the standard method for olive fruit fly control in commercial operations for the production of organic olive products. Successful mating disruption with the synthetic pheromone has also been reported[37] but has not been verified by subsequent tests. Although the major component of the blend carries only 70% of the biological activity of the blend, it is used alone in practical applications because of technical difficulties in formulating the complete blend.

E.   *BACTROCERA (DACUS) DORSALIS*

Female attraction to live males has been observed[38] and the pheromone producing glands in male flies have been identified and studied[39] but no pheromone components have been identified. Ethyl benzoate, however, has been reported as an impact ovipositional attractant.[40]

F.   *BACTROCERA (DACUS) TRYONI*

Bellas and Fletcher[41] identified six major components in the pheromone gland secretions of *D. tryoni* and *D. neohumeralis*. They found no differences among the two species and the components identified were, in order of decreasing quantity: N-3-methylbutylpropanamide, N-3-methylbutylacetamide, N-(3-methylbutyl)-2 methylpropanamide, N-2-methylbutylpropanamide, N-2-methylbutylacetamide, N-(2-methylbutyl)-2 methylpropanamide.

G.   *TOXOTRYPANA CURVICAUDA* GERSTAECKER

Chuman et al.[42] identified the pheromone of the papaya fruit fly from volatiles collected from calling males as 2-methyl-6-vinylpyrazine. The behavioral responses of the insects to the pheromone and the effects of various factors were studied[43,44] and finally the pheromone was successfully incorporated into a female annihilation scheme that still requires evaluation on large scale applications. Papaya fruit fly is only the second species (olive fly being the first) from which successful field tests of sex pheromones have been obtained.

In several other fruit fly species, especially of the genus *Bactrocera (Dacus)*, responses of virgin females to live males or male volatiles have been reported, but no chemical compounds have been identified.

IV. PARAPHEROMONES

Parapheromones are a great mystery of fruit fly biology.[45] They are very powerful male attractants eliciting pheromone-type response. The most celebrated ones are: (1) methyl-eugenol, which occurs in nature and was found to attract strongly *Dacus dorsalis, Dacus zonatus* and several other fruit fly species, (2) cue-lure and its analog zonatus's lure (only the second occurs in nature) which attract a large number of fruit flies, including *Dacus cucurbitae* and *Dacus tryoni*, and (3) the synthetic trimedlure which is a powerful attractant of *Ceratitis capitata*. Recent work indicates that a -copaene, a new plant-derived male medfly attractant could be of particular interest.[46] It was found to be much more attractive than trimedlure but only in the first few days, after which it loses its effectiveness. Latilure, a naturally occurring male attractant of *Dacus latifrons*, appears of interest for use in the management of this species which was recently introduced to Hawaii.[47]

Efforts are being concentrated on the development of controlled release formulations that are more powerful, long lasting, efficient, and of more practical use than other male lures in field applications. Trimedlure has been formulated into dispenser plugs, which prolongates the life of the attractant in nature from about 4 to more than 12 weeks.[48-51] However, further extension of lure or dispenser life is sought for cost saving in practical applications. Modifications in the molecule of known parapheromones may produce longer lasting attractants, as in the case of ceralure by McGovern and Cunningham.[52]

V. COLOR

Fruit flies locate hosts and sites for their basic feeding and reproductive activities through olfactory and visual stimuli. Color attraction is of very short range. On the other hand, it has the advantage of localized effect, it attracts both sexes almost equally and is less affected by temperature and humidity than odor attractants. Color traps, for various tephritids, are usually based either on color exclusively, or on a combination of color and olfactory attractants. It should be noted that color attraction is attractive to many insects including many beneficial species.[53]

Yellow panels and black, red, or yellow, fruit-mimicking spheres have been used most commonly as fruit fly color attractants. Colored spheres resemble fruit that may be susceptible to oviposition but, they have been found to elicit food-seeking attraction as well.[54]

Usually, visual attractants have been combined with odor attractants such as proteins, ammonium salts, pheromones or parapheromones, or combinations of more than one. One of the first patented combinations was the Pherocon trap that combined yellow color with food odors, i.e. protein hydrolysate and ammonium acetate. It has been used successfully in pest management of *Rhagoletis pomonella*.[55,56] Recently, a blend of major *Ceratitis capitata* male pheromone components was combined with black spheres for female trapping.[22]

VI. TRAPS

The attractants discussed here are sometimes combined with killing agents in adult traps, which are the tools for fruit fly monitoring, detection, and sometimes control. There are countless numbers of devices tried for fruit fly trapping. Several have never been published, and many are simply slight modifications of established traps.

The standard traps from which nearly all other trapping devices for fruit flies have evolved are the McPhail trap, the first used with liquid food baits, and Steiner or Nadel odor-attractant traps used with dry bait and various sticky traps, Jackson's being the most common, and colored sticky rectangles, e.g. Pherocon.[57] Some of these traps share common attraction or trapping elements or differ only slightly.

Besides monitoring and detection, attractants have also been applied for crop protection purposes by attracting and killing the natural population. This has been done either by spraying a food bait mixed with an insecticide or by attraction to killing devices, e.g. male annihilation[58] by attraction to a solid-carrier formulation with insecticide. These have been attempted or routinely applied extensively against fruit flies in the recent decades.[59]

For effective control, mass-trapping needs to kill a large proportion of adult fruit flies daily, because of their high reproductive rates and large natural populations.[60] The difficulty to achieve total population control is thus evident. Nevertheless, mass-trapping has been attempted and in some cases the results appear promising. In apple orchards, dark red spheres have been used to suppress *Rhagoletis pomonella* as part of an apple-pest management program.[56] In olive groves, yellow sticky panels combined with ammonium acetate dispensers gave satisfactory control of *Dacus oleae* in years of low or moderate populations.[61] Mass-trapping of *Dacus oleae* with food odor, phagostimulant, and pheromone on an insecticide treated panel produced significant population decline which reduced insecticidal applications drastically.[36]

The variety of attractants and traps often used for the same fly, necessitates the standardization and comparative evaluation of trapping systems under different climates and host situations. A recent coordinated program, organized by the Joint Division of FAO/IAEA, compared trapping of *Ceratitis capitata* with protein-hydrolysate or ammonium salt liquid bait and trimedlure dispenser in McPhail plastic and glass traps or Jackson traps.

This study was conducted in eight different countries with wide climatic and host-tree differences during a five-year period.[62-64] Although final conclusions have not been published yet, the results clearly indicate effectiveness of different trapping systems in different environments. Such collaborative programs could lead to more effective use of attractants and traps in tephritid fruit fly control or eradication programs.

REFERENCES

1. **Howlett, F. M.**, The effect of oil of citronella on two species of *Dacus*, *Transaction of the Entomological Society of London*, 60, 412, 1912.
2. **Severin, H. P. and Severin, H. C.**, A historical account on the use of kerosene to trap the Mediterranean fruit fly (*Ceratitis capitata* Wied.) *J. Econ. Entomol.*, 6, 347, 1913.
3. **Prokopy, R. J. and Roitberg, B. D.**, Foraging behaviour of true fruit flies, *American Scientist*, 72, 41, 1984.
4. **Hendrichs, J., Katsoyannos, B. I., Papaj, D. R., and Prokopy, R. J.**, Sex differences in movement between natural feeding and mating sites and tradeoffs between food consumption, mating success and predator evasion in Mediterranean fruit flies (Diptera: Tephritidae), *Oecologia*, 86, 223, 1991.
5. **Bateman, M. A. and Morton, T. C.**, The importance of ammonia in proteinaceous attractants for fruit flies family Tephritidae, *Austral. J. Agric. Res.*, 32, 883, 1981.
6. **Morton, T. C. and Bateman, M. A.**, Chemical studies on proteinaceous attractants for fruit flies, including the identification of volatile constituents, *Austral. J. Agric. Res.*, 32, 905, 1981.
7. **Mazor, M., Gothilf, S., and Galun, R.**, The role of ammonia in the attraction of females of the Mediterranean fruit fly to protein hydrolysate baits, *Entomol. Exp. Appl.*, 43, 25, 1987.
8. **Galun, R.**, Phagostimulation of the Mediterranean fruit fly, *Ceratitis capitata* by ribonucleotides and related compounds, *Entomol. Exp. Appl.*, 50, 133, 1989.
9. **Robacker, D. C., Tarshis Moreno, A. M., Garcia, J. A., and Flath, R. A.**, A novel attractant for Mexican fruit fly, *Anastrepha ludens* from fermented host fruit, *Journal of Chemical Ecology*, 16, 2799, 1990.
10. **Barki, A. and Howse, P. E.**, The influence of citrus fruit conditions on infestation by Mediterranean fruit fly, *Ceratitis capitata* (Wied.), in *Intern. Symp. Fruit Flies of Economic Importance*, Guatemala, October 1990, Abstracts, 1990, 78.
11. **Prokopy, R. J., Papaj, D. R., Hendrichs, J., and Wong, T. T. Y.**, Variation in response of *Ceratitis capitata* flies to bait spray droplets, *Entom. Exp. Appl*, In press.
12. **Drew, R. A. I. and Lloyd, A. C.**, Relationship of fruit flies (Diptera: Tephritidae) and their bacteria to host plants, *Ann. Entomol. Soc. Am.*, 80, 629, 1987.
13. **Jang, E. B. and Nishijima, K. A.**, Identification and attractancy of bacteria associated with *Dacus dorsalis* (Diptera: Tephritidae), *Environ. Entomol.*, 19, 1726, 1990.
14. **Robacker, D. C., Garcia, J. A., Martinez, A. J., and Kaufman, M. G.**, Strain of *Staphylococcus* attractive to laboratory strain *Anastrepha ludens* (Diptera: Tephritidae), *Ann. Entomol. Soc. Am.*, 84, 555, 1991.
15. **MacCollom, G. B., Lauzon, C. R., Weires, R. W., and Rutkowski, A. A.**, Attraction of adult apple maggot (Diptera: Tephritidae) to microbial isolates, *J. Econ. Entomol.*, 85, 83, 1992.

16. **Zervas, G. A.,** A new long life trap for olive fruit fly *Dacus oleae* Diptera: Tephritidae and other Diptera, *Z. Angew. Entomol.,* 94, 522, 1982.

17. **Economopoulos, A. P. and Stavropoulou-Delivoria, A.,** Yellow sticky rectangle with ammonium acetate slow-release dispenser: an efficient long-lasting trap for *Dacus oleae. Entomologia Hellenica,* 2, 17, 1984.

18. **Haniotakis, G. E. and Vassiliou-Waite, A.,** Effect of combining food and sex attractants on the capture of *Dacus oleae* flies, *Entomol. Hell.,* 5, 27, 1987.

19. **Hendrichs, J., Reyes, J., and Aluja, M.,** Behaviour of female and male Mediterranean fruit flies, *Ceratitis capitata* in and around Jackson traps placed on fruiting host trees, *Insect Sci. Applic.,* 10, 285, 1989.

20. **Yang, E. B., Light, D. M., Flath, R. A., Nagata, J. T., and Mon, T. R.,** Electroantennogram responses of the Mediterranean fruit fly, *Ceratitis capitata* to identified volatile constituents from calling males, *Entomol. Exp. Appl.,* 50, 7, 1989.

21. **Baker, P. S., Howse, P. E., Ondarza, R. N., and Reyes, J.,** Field trials of synthetic sex pheromone components of the male Mediterranean fruit fly Diptera: Tephritidae in Southern Mexico, *J. Econ. Entomol.,* 83, 2235, 1990.

22. **Heath, R. R., Landolt, P. J., Tumlinson, J. H., Chambers, D. L., Murphy, R. E., Doolittle, R. E., Dueben, B. D., Sivinski, J., and Calkins, C. O.,** Analysis, synthesis, formulation and field testing of three major components of male Mediterranean fruit fly pheromone, *J. of Chem. Ecoll.,* 17, 1925, 1991.

23. **Battiste, M. A., Strekowski, L., Vanderbilt, D. P., Visnick, M., King, R. W., and Nation, J. L.,** Anastrephin and epi-anastrephin novel lactone components isolated from the sex pheromone blend of male Caribbean and Mexican fruit flies *Anastrepha spp., Tetrahedron Lett.,* 24, 2611, 1983.

24. **Robacker, D. C. and Heath, W. G.,** (Z)-3-nonenol, (Z,Z)-3,6 nonadienol and (S,S)-(+)-epianastrephin male produced phoromones of the Mexican fruit fly, *Anastrepha ludens, Entomol. Exp. Appl.,* 39(2), 103, 1985.

25. **Robacker, D. C. and Wolfenbarger, D. A.,** Attraction of laboratory reared irradiated Mexican fruit flies to male produced pheromone in the field, *Southwest. Entomol.,* 13, 75, 1988.

26. **Robacker, D. C.,** Behavioral responses of female Mexican fruit fly *Anastrepha ludens* to components of male-produced sex pheromones, *J. Chem. Ecol.,* 14, 1715, 1988.

27. **Robacker, D. C. and Garcia, J. A.,** Responses of laboratory-strain Mexican fruit flies, *Anastrepha ludens,* to combinations of fermenting fruit odor and male produced pheromone in laboratory bioassays, *J. Chem. Ecol.,* 16, 2027, 1990.

28. **Robacker, D. C., Garcia, J. A., and Hart, W. G.,** Attraction of a laboratory strain of *Anastrepha ludens,* Diptera: Tephritidae to the odor of fermented chapote fruit and to pheromones in laboratory experiments, *Environ. Entomol.,* 19, 403, 1991.

29. **Heath and Battiste,** Unpublished report to California Dept. of Food and Agriculture, 1991.

30. **Nation, J. L.,** The sex pheromone blend of Caribbean fruit fly males: Isolation, biological activity, and partial chemical characterization, *Environ. Entomol.,* 4, 27, 1975.

31. **Nation, J. L.,** Biology of pheromone release by male Caribbean fruit flies, *Anastrepha suspensa,* Diptera: Tephritidae, *J. Chem. Ecol.,* 16, 553, 1990.

32. **Mazomenos, B. E. and Haniotakis, G. E.,** Male olive fruit fly, *Dacus oleae* attraction to synthetic sex pheromone components in laboratory and field tests, *J. Chem. Ecol.,* 11, 397, 1985.

33. **Haniotakis, G., Francke, W., Mori, K., Redlich, H., and Schurig, V.,** Sex specific activity of R-(-) and S-(+)-1,7 dioxaspiro [5.5] undecane, the major pheromone of *Dacus oleae., J. Chem. Ecol.,* 12, 1559, 1986.

34. Niccoli, A. and Tiberi, R., The biological activity of *Dacus oleae* and *Prays oleae* sexual attractants: the relationship between trap captures and infestation, *Redia*, 65, 407, 1982.

35. Haniotakis, G. E., unpublished data, 1992.

36. Haniotakis, G., Kozyrakis, M., Fitsakis, T., and Antonidaki A., An effective mass trapping method for the control of *Dacus oleae* (Diptera:Tephritidae), *J. Econ. Entomol.*, 84, 564, 1991.

37. Jones, O. T., Lisk, J. C., Howse, P. E., Baker, R., Bueno, A. M., and Ramos, P., Mating disruption of the olive fruit fly *Dacus oleae* with the major component of its sex pheromone, in *Proc. of the CEC/IOBC Intern. Symposium, Athens, Greece, Nov. 16-19, 1982, Fruit Flies of Economic Importance XII*, Cavalloro, R., Ed., A. Balkema, Rotterdam, 1983, 500.

38. Kobayashi, R. M., Ohinata, K., Chambers, D. L., and Fujimoto, M. S., Sex pheromones of the Oriental fruit fly and the melon fly mating behaviour bioassay method and attraction of females by live males and by suspected pheromone glands of males, *Environ. Entomol.*, 7, 107, 1978.

39. Lee, W. Y. and Chang, T. H., Studies on the development of sex pheromone producing gland of the male oriental fruit fly *Dacus dorsalis* Hendel with scanning Electron Microscopy, *Chin. J. Entomol.*, 9, 69, 1989.

40. Chiu, H. T., Ethyl benzoate an impact ovipositional attractant of the Oriental fruit fly, *Dacus dorsalis*, Hendel, *Chin. J. Entomol.*, 10, 375, 1990.

41. Bellas, T. E. and Fletcher, B. S., Identification of the major components in the secretion from the rectal pheromone glands of the Queensland fruit flies, *Dacus tryoni* and *Dacus neohumeralis*, Diptera: Tephritidae, *J. Chem. Ecol.*, 5, 795, 1979.

42. Chuman, T., Landolt, P. J., Heath, R. R., and Tumlinson, J. H., Isolation, identification, and synthesis of male produced sex pheromone of papaya fruit fly, *Toxotrypana curvicauda* Gerstaecker, Diptera: Tephritidae, *J. Chem. Ecol.*, 13, 1979, 1987.

43. Landolt, P. J. and Heath, R. R., Effects of age, mating, and the time of day on behavioral responses of female papaya fruit fly *Toxotrypana curvicauda* Gerstaecker Diptera: Tephritidae to synthetic sex pheromone, *Environ. Entomol.*, 17, 47, 1988.

44. Landolt, P. J. and Heath, R. R., Effects of pheromone release rate and time of day on catches of male and female papaya fruit flies, Diptera: Tephritidae on fruit model traps baited with pheromone, *J. Econ. Entomol.*, 83, 2041, 1990.

45. Cunningham, R. T., Parapheromones, in *Fruit Flies, Their Biology, Natural Enemies and Control*, Vol. 3A, Robinson, A. S. and Hooper, G., Eds., Elsevier, Amsterdam, 1989, 221.

46. Flath, B. and Teranishi, R., personal communication, 1991.

47. Carey, J. R., personal communication, 1991.

48. McGovern, T. P., Cunningham, R. T., and Leonhardt, B. A., Crystallization inhibitors and extenders for trimedlure, the attractant for the Mediterranean fruit fly (Diptera: Tephritidae), *J. Econ. Entomol.*, 80, 806, 1987.

49. Leonhardt, B. A., Cunningham, R. T., Rice, R. E., Harte, E. M. and McGovern, T. P., Performance of controlled-release formulations of trimedlure to attract the Mediterranean fruit fly, *Ceratitis capitata*, *Entomol. Exp. Appl.*, 44, 45, 1987.

50. Leonhardt, B. A., Cunningham, R. T., Rice, R. E., Harte, E. M., and Hendrichs, J., Design, effectiveness, and performance criteria of dispenser formulations of trimedlure, an attractant of the Mediterranean fruit fly (Diptera: Tephritidae,) *J. Econ. Entomol.*, 82, 860, 1989.

51. Baker, P. S., Hendrichs, J., and Liedo, P., Improvement of attractant dispensing systems for the Mediterranean fruit fly Diptera, Tephritidae sterile release program in Chiapas Mexico, *J. Econ. Entomol.*, 81, 1068, 1988.

52. **Wood, M.,** New lure, attracts medfly males, *Agricultural Research*, Feb., 19, 1989.

53. **Economopoulos, A. P.,** Use of traps based on colour and/or shape, in *Fruit Flies, Their Biology, Natural Enemies and Control*, Vol. 3B, Robinson, A. S. and Hooper, G. Eds., Elsevier, Amsterdam, 1989, 315.

54. **Katsoyannos, B. I.,** Field responses of Mediterranean fruit flies to coloured spheres suspended in fig, citrus and olive trees, in *Insects - Plants*, Labeyrie, V., Fabres, G., and Lachaise, D., Eds., Dr. W. Junk Publishers, Dordrecht, The Netherlands, 1987, 167.

55. **Neilson, W. T. A., Rivard, I., Trottier, R., and Whitman, R. J.,** Pherocon AM standard traps and their use to determine spray dates for control of the apple maggot, *J. Econ. Entomol.*, 69, 527, 1976.

56. **Prokopy, R. J.,** A low -spray apple-pest-management program for small orchards, *Can. Ent.*, 117, 581, 1985.

57. **Cunningham, R. T.,** Population detection, in *Fruit Flies, Their Biology, Natural Enemies and Control*, Vol. 3B, Robinson, A. S. and Hooper, G. Eds., Elsevier, Amsterdam, 1989, 169.

58. **Cunningham, R. T.,** Male annihilation, in *Fruit Flies, Their Biology, Natural Enemies and Control*, Vol. 3B, Robinson, A. S. and Hooper, G. Eds., Elsevier, Amsterdam, 1989, 345.

59. **Roessler, Y.,** Insecticidal bait and cover sprays, in *Fruit Flies, Their Biology, Natural Enemies and Control*, Vol. 3B, Robinson, A. S. and Hooper, G. Eds., Elsevier, Amsterdam, 1989, 329.

60. **Weidhaas, D. E. and Haile, D. G.,** A theoretical model to determine the degree of trapping required for insect population control, *E.S.A. Bulletin*, 24, 18, 1978.

61. **Economopoulos, A. P., Raptis, A., Stavropoulou-Delivoria, A., and Papadopoulos, A.,** Control of *Dacus oleae* by yellow sticky traps combined with ammonium acetate slow-release dispensers, *Entomol. Exp. Appl.*, 41, 11, 1986.

62. **International Atomic Energy Agency,** Standardization of medfly trapping for use in sterile insect technique programmes, Report of coordination meeting, Joint FAO/IAEA Division, IAEA, Vienna, Austria, 1987.

63. **International Atomic Energy Agency,** Standardization of medfly trapping for use in sterile insect technique programmes. Report of coordination meeting, Joint FAO/IAEA Division, IAEA, Vienna, Austria, 1988.

64. **International Atomic Energy Agency,** Standardization of medfly trapping for use in sterile insect technique programmes. Report of coordination meeting, Joint FAO/IAEA Division, IAEA, Vienna, Austria, 1990.

Chapter 5

MASS REARING OF FRUIT FLIES:
A DEMOGRAPHIC ANALYSIS

Pablo Liedo, James R. Carey, and Roger I. Vargas

I. INTRODUCTION

The ability to produce large numbers of insects at a relatively low cost has long been recognized as a major requirement for the successful application of the Sterile Insect Technique (SIT). During the past three decades, technologies have been developed to mass produce, at levels up to one billion flies per week, some of the most economically important species of fruit flies: the Mediterranean fruit fly, *Ceratitis capitata*; the melon fly *Bactrocera cucurbitae*; the oriental fruit fly *B. dorsalis*.[1-3]

Most of the research to support large rearing facilities during recent years has been devoted to: (1) reduce production costs through recycling larval diets and developing new diets with cheaper ingredients;[4-8] (2) develop methods to assess and improve quality of both production process and insects produced;[9-13] and (3) evaluate genetic sexing strains under mass rearing conditions.[14]

An alternative to optimize the production process is the application of population harvesting demographic models. Carey and Vargas[15] proposed such a model for insect rearing. In this paper we review the results from applying this model to six species of tephritid fruit flies and discuss the implications of these results.

II. THE MODEL

In a mass rearing facility, a decision must be made on how many individuals are needed for colony maintenance. To optimize the process, this fraction of the population needs to be the minimum possible to maintain production at the desired level. This reduces production costs and stress to the colony due to unnecessary crowding. Therefore, a fruit fly population in a mass rearing facility is a population with a growth rate of zero. Harvesting of pupae and discarding of old adults are considered as mortality factors for the model. The age structure of this population is determined by the particular harvesting and rearing strategy. Trade offs between the fraction that is harvested and the adult discard age are analyzed to identify the combination that yield the largest pupal production per female. The demographic formulae and symbols for the model are summarized in Table 1.

Table 1

Demographic Formulae and Notation of the Demographic Parameters Used in the Mass Rearing Model[15]

Symbol	Parameter	Formula
R_d	Net reproduction	$\sum\limits_{x=a}^{d} l_x m_x$
h	Fraction of the population that is harvested.	$1-(R_d)^{-1}$
	Fraction of the population that is used for replacement in the colony	$1-h$
S	Sum of all individuals alive at the facility (excluding those harvested)	$\sum\limits_{x=0}^{0-1} l_x + (1-h)\sum\limits_{x=0}^{d} l_x$
C_x	Fraction of the standing population at age x in a Stable Age Distribution	l_x/S for $x < \theta$ $l_x(1-h)/S$ for $x > \theta$
P	Daily pupal production per female	$2l_\theta (R_d - 1)/\sum\limits_{x=E}^{d} l_x$
R	Total number of individuals in the released population ($N=$ number of females in the colony)	$PN\sum\limits_{x=E}^{w} l_x R$
C_x^R	Fraction at age x in the released standing population	$l_x/\sum\limits_{x=E}^{w} l_x^R$

Notation: $x=$ age in days; $\theta=$ age at the harvesting stage (mature pupae); $E=$ age at adult eclosion; $a=$ age at first reproduction; $d=$ adult discard age; $w=$ last possible day of life; $l_x=$ probability of surviving to age x; $m_x=$ number of eggs produced per female at age x; $l_x R =$ probability of surviving to age x in the released population.

III. APPLICATION TO FRUIT FLIES

Carey and Vargas[15] applied the model to data from Vargas et al.[16] related to the three laboratory reared tephritids in Hawaii: *C. capitata, B. dorsalis,* and *B. cucurbitae.* Liedo and Carey[17] applied the model to three species of *Anastrepha* fruit flies of economic importance: *A. ludens, A. obliqua,* and *A. serpentina*; determining their survival and fecundity rates from wild flies obtained from field collected infested fruits.

The optimal discard age, harvesting fraction, production rate and net fecundity determined by the model for these six species are shown in Table

2. For comparison, we also include the life time net fecundity rates that have been reported for these species.[18-21]

A significant difference in the optimal discard age and production rate can be observed when comparing the three *Anastrepha* species with the other three species. A reason for this is that the survival and fecundity data for the *Anastrepha* group were obtained from wild flies, whereas the other species were laboratory reared strains. The mass rearing process has shown to select for early and high reproduction. For example, when comparing reproductive parameters of laboratory and wild *B. dorsalis*, Foote and Carey[19] found that fecundity of the laboratory strain was 10-fold greater (120 vs. 1236 eggs/female for wild and lab strains respectively). In *C. capitata*, Vargas and Carey[20] reported a total production of 1068 eggs per female in a mass reared strain and from 485 to 930 eggs per female in four wild strains.

These differences in reproductive traits could be the result of unconscious selection during colonization or adaptation of wild strains to mass rearing conditions, since usually the eggs collected at the earliest ages are the ones that form the next generation. Those produced by old females are discarded.[22] This selection favors early reproduction.

Although there are significant differences in discard age and female production, the fraction to be harvested in all six species only ranged from 98.59 to 99.49%.

The age distribution at the rearing facility for each of the six species is shown in Table 3. Note that the fraction at the adult stage represents in all species less than 1%. The fraction at the larval and pupal stages ranged from 78 to 91% of the total population.

The model can also be used to determine the number of individuals in the released population and their age distribution (Table 1). Liedo and Carey[17] estimated this for the *Anastrepha* species, assuming same survival rates as those obtained in the laboratory experiments.

IV. GENERAL DISCUSSION

Mass rearing still is the most expensive activity of most SIT programs. The production of the maximum possible number of sterile flies at the lowest possible cost requires optimization of many aspects at the facility. These population harvesting demographic models provide elements for such optimization.

Current programs can determine their optimal parameters according to the characteristics of the flies they are rearing. From the model, managers will be able to determine if changes in the time for discard or in the fraction harvested are required.

Table 2
Optimal Discard Age in Days (d), Daily Pupal Production Per Female (P), Percentage to be Harvested (h), Net Fecundity at Discard Age in Female Eggs per Female (R_d), and Life Time Net Fecundity in Female Eggs per Female of Six Species of Tephritid Fruit Flies

Species	Discard Age d	Production Rate P	Fraction Harvested h	Net Fecun. R_d	Life Time Net Fecun. R_o
C. capitata[15,20]	14	21	99.30	142	1068
B. dorsalis[15,19]	25	20	99.49	198	1236
B. cucurbitae[15,18]	20	9	98.96	96	709
A. ludens[8,17]	64	4	99.21	126	597
A. obliqua[8,17]	41	4	98.66	74	592
A. serpentina[8,17]	36	3	98.59	71	244

Table 3
Percentage of Standing Population of Six Species of Fruit Flies in Four Stages at a Mass Rearing Facility (Exclusive of Harvested Pupae) Using Optimal Discard Ages[15,17]

Species	Eggs	Larvae	Pupae	Adults
C. capitata	12.1	40.2	47.2	0.5
B. dorsalis	13.6	41.4	44.5	0.5
B. cucurbitae	7.0	42.6	49.5	0.9
A. ludems	16.2	50.6	32.7	0.5
A. obliqua	20.8	43.9	34.9	0.4
A. serpentina	18.8	38.1	42.4	0.7

For new programs, results from the model can help in two ways: (1) by providing elements for the facility design in terms of number of cohorts needed, and area required for each stage according to the level of production desired, and (2) by determining when would be the best time to discard the adults and what would be the fraction to be harvested.

For colonization or adaptation of wild flies to mass production conditions, this model helps to determine in advance the optimal discard age and to develop strategies that can be added to those that consider genetic diversity by collecting representative samples from field populations.[23,24] By maintaining wild flies until the optimal discard age, we maximize their egg production and avoid detrimental selection of some important traits that could be associated with late reproduction or old ages, such as longevity.[3,22]

The selection for early and high reproduction might be required to optimize mass production. However, it could result in harmful selection against some other desirable traits. The optimal age to discard ovipositing females, determined by the model with data from wild flies, should help to avoid this harmful selection.

Finally, if survival rates and the reproductive life of the sterile flies are known, this demographic model can tell how many insects we need to produce in order to have a given field population of sexually active flies. Certainly, this will make the SIT more predictable and will reduce the approach of release as many flies as possible until control is achieved.

ACKNOWLEDGMENTS

We acknowledge the financial support from the Consejo Nacional de Ciencia y Tecnologia (CONACYT), Mexico, grant A128ccde-920281(CN-7), and the Organizing Committee of the XIX International Congress of Entomology.

REFERENCES

1. **Schwarz, A. J., Zambada, A., Orozco, D. H. S, Zavala, J. L., and Calkins, C. O.,** Mass production of the Mediterranean fruit fly at Metapa, Mexico, *Fla. Entomol.,* 68: 467, 1985.

2. **Vargas, R. I.,** Mass production of Tephritid fruit flies, in *Fruit Flies: Their Biology, Natural Enemies and Control,* Vol. 3B, Robinson A. S. and Hooper, G., Eds., Elsevier, Amsterdam, 1989, 141.

3. **Kakinohana, H. and Yamagishi, M.,** The mass production of the melon fly - techniques and problems, in *Proc. Intl. Symp. on the Biology and Control of Fruit Flies, Okinawa, Japan,* Kawasaki, K., Iwahashi, O., Kaneshiro, K. Y., Eds., 1991, 1.

4. **Bruzzone, N. D. and Schwarz, A.,** Recycling larval rearing medium for Mediterranean fruit fly mass production: A preliminary experiment, *J. Appl. Entomol.,* 103, 418, 1987.

5. **Bruzzone, N. D., Economopoulos, A., Andrade, L., Caceres, C., and Rendon, P.,** Technology transfer and medfly mass-production, in *Fruit Flies: Biology and Management,* Aluja, M. and Liedo, P., Eds., Springer-Verlag, New York, 1993, 249, Part VIII.

6. **Dominguez, J., Zavala, J. L., Liedo, P., and Bruzzone, N.,** Implementation of the starter diet technique for medfly mass-rearing at Metapa, Chiapas, Mexico, in *Fruit Flies: Biology and Management,* Aluja, M. and Liedo, P., Eds., Springer-Verlag, New York, 1992, 277. (In press).

7. **Economopoulos, A. P., Al-Taweel, A. A., and Bruzzone, N. D.,** Larval diet with starter phase for mass-rearing *Ceratitis capitata*: substitution and refinement in the use of yeasts and sugars, *Entomol. Exp. Appl.,* 55, 239, 1992.

8. **Liedo, P., Zavala, J. L., Orozco, D., Fredersdorff, C., and Schwarz, A. J.,** Ten years of successful sterile medfly mass-production at Metapa, Chiapas, Mexico, in *Fruit Flies: Biology and Management,* Aluja, M. and Liedo, P., Eds., Springer-Verlag, New York, 1993, 269 , Part VII.

9.   **Calkins, C. O.**, Quality control, in *Fruit Flies: Their Biology, Natural Enemies and Control,* Vol. 3B, Robinson A. S. and Hooper, G., Eds., Elsevier, Amsterdam, 1989, 153.

10.   **Calkins, C. O. and Ashley, T. R.**, The impact of poor quality of mass reared Mediterranean fruit flies in the SIT used for eradication, *J. Appl. Entomol.,* 108, 401, 1989.

11.   **Leppla, N. C. and Ashley, T. R.**, Quality control in insect mass production: a review and model, *Bull. Entomol. Soc. Am.,* 35, 33, 1989.

12.   **Kaneshiro, K. Y.**, Quality control of mass-reared strains of fruit flies, in *Proc. Intl. Symp. on the Biology and Control of Fruit Flies, Okinawa, Japan,* Kawasaki, K., Iwahashi, O., Kaneshiro, K. Y., Eds., 1991, 138.

13.   **Bruzzone, N., Caceres, C., Andrade, L., Guzman, N., Calderon, J., and Rendon, P.,** Process control for Medfly mass production at San Miguel Petapa, Guatemala: a system approach, in *Fruit Flies: Biology and Management,* Aluja, M. and Liedo, P., Eds., Springer-Verlag, New York, 1993, 289 , Part VIII.

14.   **Economopoulos, A. P., Nitzan, Y., and Rossler, Y.**, Mass rearing, quality control and male only sterile insect technique application with a pupa color genetic sexing strain of the Mediterranean fruit fly, in *Fruit Flies: Biology and Management,* Aluja, M. and Liedo, P., Eds., Springer-Verlag, New York, 1992, 267 (In press).

15.   **Carey, J. R. and Vargas, R. I.**, Demographic analysis of insect mass rearing: a case study of three tephritids, *J. Econ. Entomol.,* 78, 523, 1985.

16.   **Vargas, R. I., Miyashita, D., and Nishida, T.**, Life history and demographic parameters of three laboratory-reared tephritids, *Ann. Entomol. Soc. Am.,* 77, 651, 1984.

17.   **Liedo, P. and Carey, J. R.**, Mass rearing of *Anastrepha* fruit flies: a demographic analysis, *J. Econ. Entomol.,* 1993 (Submitted).

18.   **Carey, J. R., Harris, E. J., and McInnis, D. O.**, Demography of a native strain of the melon fly, *Dacus cucurbitae* Coquillet, from Hawaii, *Entomol. Exp. Appl.,* 38, 195, 1985.

19.   **Foote, D. and Carey, J. R.**, Comparative demography of a laboratory and wild strain of the oriental fruit fly, *Dacus dorsalis, Entomol. Exp. Appl.,* 44, 263, 1987.

20.   **Vargas, R. I. and Carey, J. R.**, Comparison of demographic parameters for wild and laboratory-adapted Mediterranean fruit fly (Diptera: Tephritidae), *Ann. Entomol. Soc. Am.,* 82, 55, 1989.

21.   **Liedo, P., Carey, J. R., Celedonio, H., and Guillen, J.**, Size specific demography of three species of Anastrepha fruit flies, *Entomol. Exp. Appl.,* 1993 (In press).

22.   **Economopoulos, A. P.**, Adaptation of the Mediterranean fruit fly (Diptera: Tephritidae) to artificial rearing, *J. Econ. Entomol.,* 85, 753, 1992.

23.   **Leppla, N. C. and Ozaki, E.**, Introduction of a wild strain and mass rearing of medfly, in *Proc. Intl. Symp. on the Biology and Control of Fruit Flies, Okinawa, Japan,* Kawasaki, K., Iwahashi, O., Kaneshiro, K. Y., Eds., 1991, 148.

24.   **Bartlett, A. C.**, Guidelines for genetic diversity in laboratory colony establishment and maintenance, in *Handbook of Insect Rearing.* Singh, P. and Moore, R. F., Eds., Elsevier, Amsterdam, 1984, 7.

Chapter 6

NUTRITIONAL, BIOCHEMICAL, AND BIOLOGICAL ASPECTS OF QUALITY CONTROL IN THE OLIVE FRUIT FLY

A. G. Manoukas

I. INTRODUCTION

It is generally accepted that mass rearing of insects leads eventually, to the development of an insect type that differs from the source population. This was a concern when we began to colonize the olive fruit fly *Dacus oleae* (Gmelin) (Diptera: Tephritidae) for control. Several colonies of the olive fruit fly were established in our laboratory during the last 25 years to study and control this insect with various methods, including the Sterile Insect Technique (SIT). Application of this technique on a small scale failed to control the natural population, however. Quality of the released flies may have been responsible for the failure.

It is well known that insect quality and its control is very important for SIT and depends upon the quality of the products and procedures involved in the production system. The nutritional quality of diet is of great importance in this system because it provides energy and all essential nutrients for insect performance. The nutrients that must be supplied in food are essentially the same in all insects because, in general, the metabolic machine is remarkably uniform among and between species. Of course, genetic variations in minor nutrient requirements do occur, but these are generally of a quantitative rather than qualitative nature. The quantitative requirement of the artificially reared strains of the olive fruit fly for the various nutrients and energy is unknown and this makes the association of a particular nutrient and of a given insect quality very difficult.

The nutritional profile of the artificial diet (especially larval diet) used for the olive fruit fly was found to be quite different, in certain respects, than that of the natural food. Differences in body composition and behavior were also observed between wild and artificially reared insects. The purpose here is to present the nutritional differences that occur between natural food and artificial diet, and the differences found in quality characteristics of the wild type (WT) and laboratory type (LT) fly and certain aspects of nutrition that may be related to insect quality. Differences were mostly those observed or studied in our laboratory.

II. NUTRITIONAL DIFFERENCES IN THE FOOD OF WILD AND ARTIFICIALLY REARED OLIVE FRUIT FLY

In nature, the adult olive fruit fly feeds on various mostly unknown natural products and the larvae feed exclusively in the olive fruit mesocarp.

0-8493-4854-4/94/$0.00+$.50

Recently, the artificial rearing of the olive fruit fly has been reviewed by Tzanakakis[1] and the feeding and nutrition in nature and in the laboratory was reviewed by Tsitsipis.[2]

A.   ADULT FOOD

Although the nutritional profile of adult food in the laboratory is quite well known, no comparison could be made with the diet of wild flies because the food consumed by the wild adults in nature is not known. Adults held in the laboratory were able to survive and reproduce on aphid honeydew or on pollen when fed with a sucrose solution, but they exhibited inferior behavior when compared with those fed the artificial adult diet.[3]

B.   LARVAL FOOD

The nutritional composition of the larval food of the olive fruit fly in nature and in the laboratory has been reported elsewhere.[4-6] Comparison of the composition of the two types of food showed certain differences that may be related to quality differences in wild and laboratory insects. Energy to protein ratio reported for the artificial diet as 0.11 is very low compared to that for the olive fruit mesocarp which is 0.8 when the olive fruit contains about 4% lipids at early maturity. This is the stage when the olive fruits are susceptible to larval attack, but critical components of the olive fruit for oviposition or larval growth are unknown. Also at this stage the Non-Protein Nitrogen (NPN) content of the olive fruit is at a very low level.[4] Unfortunately, no work is known of the role of NPN or of any other factor upon larval growth and development other than that speculated by Sawas-Dimopoulou and Fytizas who reported on the role of free amino-acids. Hagen[8] did report on the role of symbiotes on the assimilation of NPN, however. In addition, the energy to NPN ratio is about 100 times lower in the artificial diet than in the olive fruit. Among the individual amino acids, total threonine and lysine is about 8 times higher, and total methionine is about 4 times higher in the artificial diet than in the olive fruit. The lysine to arginine ratio is 0.5 in the olive fruit and 1.7 in the artificial diet. Here an antagonism between lysine and arginine appears to be well documented in chicks in which the ratio of dietary lysine to arginine cannot be greater than 1 before growth retardation can occur with small additional amounts of lysine.[9] The lysine to arginine ratio in the olive fruit is well below 1 and that of the diet 1.7. When lysine was added at 0.4 g/100 ml of diet, a growth retardation occurred, probably because the ratio became higher in the diet. On the contrary, it took a large amount of arginine (1.6 g/100) before growth retardation could be observed.[10] Other amino acid toxicities such as methionine may be involved in the quality of this insect.[11]

Because of the differences in mineral compositions of the diet, the balance of certain of these minerals is quite different between the two foods.

The ratio of K/Na is 143 for the olive fruit and only 2.6 for the artificial diet. The ratio of P/Ca is also quite different, that is, 51 for the olive fruit and 182 for the artificial diet.[12] Finally, copper is very high in the artificial diet compared to the olive fruit, and it may be detrimental to the artificially reared larva. Toxicity of copper has been ascribed to interfere with enzymatic processes or to chelate with amino acids making them unavailable to the organism.[13] Excess copper also seems to influence the availability of fatty acids in the insect tea totrix.[14] In addition, while screening a large number of ingredients for inclusion in a diet for mass rearing the olive fruit fly, it was found that when total copper in the diet was higher than 1.2 mg/100 g a growth retardation was observed.[15] Consideration of these data suggests that nutritional imbalances, antagonisms, or toxicities may play an important role upon many of the quality problems reported for the artificially reared flies. Further studies on the nutrition of the olive fruit fly and the biochemistry of olive fruit and the artificial diet are necessary to substantiate and solve the related biochemical and biological differences reported in this paper.

Olive fruit was found to be rich in tocopherols and tocotrienols[16] compared to artificial larval food which contains about four times less tocopherols.[17] The olive fruit is very rich in phenolics and in many other chemicals which may have a nutritional, antioxidant, and phagostimulatory action. The artificial diet, on the other hand, contains preservatives proven to be detrimental to egg hatchability.[18] Almost nothing is known on the effects of non-nutritional factors (texture, allelochemicals etc.) of the two types of food upon the insect.

III. BIOCHEMICAL DIFFERENCES IN WILD TYPE AND LABORATORY TYPE OLIVE FRUIT FLY

The entire make-up of an insect body is controlled by genes received from its parents. These include not only physical qualities that are readily observed, but also metabolic characteristics that include enzyme systems responsible for the ability to utilize nutrients and synthesize essential body metabolites.

A. ENZYME SYSTEMS

Certain enzymes of wild type (WT) and laboratory type (LT) olive fruit flies have been determined electrophoretically by Bush and Kitto.[19] The biochemical genetics and the enzyme systems of the olive fruit fly have been reviewed and discussed recently by Zouros and Loukas.[20] The comparison of the enzyme systems of WT and LT insects has demonstrated the action of selection during the critical period of adaptation to artificial rearing. The evidence suggested for the alcohol dehydrogenase (ADH), is that the allele gene frequency change is due to the artificial diet. This change occurred within four generations of artificial rearing.[21] However, when LT flies oviposited in olive fruits, gene frequency at ADH locus returned to the pattern of wild flies in the first generation.[21] For example, gene frequency of allele 1 (Slow) changed from 0.6 in WT to 0.2 in LT insects, and it came back to 0.5 when LT flies oviposited in olives.

B. PROTEIN AND AMINO ACIDS

Significant differences were found in total protein between wild and laboratory larvae. Despite the constancy of whole body and protein composition for most amino acids in both types of insects, two amino acids (tyrosine and proline) and three amino acids (tyrosine, valine, and glutamic acid) differed in the two types of adults and three amino acids in two types of larvae.[22] Significant changes were found between the two types of insects and the two stages of development (larvae vs. adult) in free and peptide-bound amino acids.[23] For example, WT larvae contained 35.2 and 16.0 µmoles/g wet weight free and peptide bound amino acids, while the respective figures were 25.3 and 31.3 for the LT larvae. Free amino acids were 54.4 and 24.1 µmoles for WT and LT adults respectively. The free amino acid pattern of LT larvae was considerably different from that of WT larvae, while that of laboratory adults was similar to wild adults, to most amino acids. Thus, WT and LT larvae were found to differ significantly in nine amino acids (histidine, tyrosine, leucine, isoleucine, caline, aspartic acid, glycine, and proline), while WT and LT adults differ in only four amino acids (lysine, methionine, tyrosine, and alanine). In addition, the ratio of essential to non-essential amino acids was 1.0 for WT larvae and adults, while this was 0.8 and 0.6 for LT larvae and adults, respectively. The peptide bound amino acid content in WT larvae was approximately 50% of that found in LT larvae, while WT and LT adults did not differ. On the contrary, free amino acid content was higher in WT insects compared to LT insects. In addition, the ratio of free to peptide-bound amino acid content for LT larvae was 0.8 and LT adults 1.3, as opposed to 2.2 and 2.1 for WT larvae and adults. This may have imposed a stress upon the LT insects that had smaller body weight and lower total nitrogen content. It should also be noted, that five essential amino acids were found to be absent or in trace amounts in WT larvae peptides, compared to only two found absent in LT insects.

C. FATS AND LIPID CONSTITUENTS

Significantly different values were given for certain lipid fractions of WT and LT adults. Young females on artificial diet had more acidic phospholipids than young females reared on olives.[24] Specifically, these phospholipids were 3.9% of the total lipids for WT, 7h old unfed females and 7.2% for LT females.

D.   PHEROMONES

Differences were found in pheromone production between WT and LT flies, under certain environmental conditions. Haniotakis[25] reported that LT flies, fed artificial diet, contained more pheromone (127.8 ng/female) in the Z2b fraction, which is the main component, than WT flies (44.2 ng/female). This may explain why LT females were more attractive to both LT and WT males in laboratory and field studies. In a 19 test experiment, as already mentioned, the LT insects fed the artificial diet produced 127.8 ng/female Z2b, while the LT insects fed only water sucrose solution produced 99.0 ng/female. Contrary to these findings, Mazomenos and Mazomenou[26] reported that rectal gland extracts collected from WT females taken from olives as pupae and fed the artificial diet after emergence produced more pheromone (1.1+0.0 μg/female) compared to LT adults (0.3+0.0 μg/female).

E.   EYE PIGMENTATION

The eye pigmentation differences observed between WT and LT insects[27] may be due to chemical differences between the natural and artificial food of this insect.

F.   ASH AND INORGANIC ELEMENTS

Despite the fact that ash content was not significantly different between WT and LT newly emerged adults, several differences were determined in individual trace elements. Thus WT and LT flies differed significantly in Mn, Cu, Cr, and Zn.[28]

IV. BIOLOGICAL DIFFERENCES IN WILD AND LABORATORY REARED OLIVE FRUIT FLIES

WT flies (reared as larvae in olive fruit) and LT flies (artificially reared for over 6 generations) were fed the adult artificial diet, oviposited in artificial devices, and kept under same laboratory conditions, unless otherwise indicated.

A.   HATCHABILITY, PUPAL AND ADULT WEIGHT

Hatchability of eggs obtained from WT adults laid on oviposition parafine devices was consistently lower (but usually not significantly lower) than hatchability of eggs obtained from WT adults laid in olives or from well adapted (over 6 generations) LT insects. The same trend was observed for pupal and adult weight, but in this case LT insects usually exhibited significantly lower values than WT insects. In nature, pupal and adult weights varied greatly; but under laboratory conditions, pupal weight from larvae reared in olives averaged 6.6+0.8 mg as compared to 5.5+0.7 for LT pupae reared for six generations at a density of 4.1 neonate larvae/g diet.[29] Of course, in artificial rearing, pupal and adult weights depend upon larval density and varied between 1.9 and 5.9 mg for larval density ranging from 68 to 8 neonate larvae/g diet.[30] Larval survival up to the 5th day was not affected by larval density, but total larval developmental period and survival to pupation was greatly affected.[31] Thus, pupation occurred on the 11th, 12th, 13th, and 14th days for density 10, 20, 40, and 60 eggs/g of diet, respectively, and survival to pupation was 72, 68, 55, and 40% of neonate larvae for the above egg densities, respectively. These findings suggest that, in order to improve survival and pupal weight, different diets should be used for different phases of larval growth and development as routinely practiced in higher animals or nutritional supplements should be added to the same diets.[12]

B.   ADULT EMERGENCE, SURVIVAL, AND REPRODUCTION

Comparisons of adult emergence, survival, and reproduction were made between WT and LT flies under laboratory conditions while feeding both types of flies the adult artificial diet. There were great differences in average figures reported due to many factors involved, but several means reported may have been very critical for SIT outcome. Economopoulos and Tzanakakis[32] found that WT flies lived up to 7-8 weeks while LT adults only lived 4-5 weeks. With improved dietary and rearing conditions, however, this difference may be diminished. Heavier LT adults (of 5.9 mg/insect) only lived 19 days.[33] In another test, the average life was 53 days for WT adults and 36 days for LT adults under standard laboratory conditions.[34] In a test involving survival without food and water (but under different RH conditions), WT flies lived twice as long as LT flies (4.4 vs. 2.8 days). Differences were smaller when flies of the same mean weight were used (3.2 days vs. 2.8 days). When only water was available, WT flies of 6.8 mg pupal weight lived an average of 5.2 days while LT flies lived 4.3 days. It is suggested that survival with or without water be included as a quality test. LT flies survived for two to three weeks when released in the field.[35]

Genetic selection in the laboratory has resulted in a more rapid sexual maturity rate for LT flies. Wild flies (males and females) were sexually mature on the 8th day of their life, while maturity of LT flies occurred in the second day for males and on the third day for females. Similarly, WT flies were ready to oviposit seven to nine days after emergence and LT flies when they were four to five days old.[36] There was a definite effect of larval density associated with different pupal and adult weight on the number of eggs produced by the corresponding females.[33] Flies reared at a density of nine neonate larvae/g of larval diet developed into 6.3 mg female pupae, laid 17 eggs/female/day while females reared at densities of 54 neonate larvae /g diet resulting in 4.5 mg females that laid nine eggs/female/day. All adults were fed the same artificial diet. This diet fed during the adult stage did not compensate in egg production for the different dietary stresses applied during the larval stage.[30]

C. FLIGHT ABILITY, DISPERSAL, AND RESPONSE TO COLOR

WT flies showed a greater tendency to disperse and a greater ability to fly than LT flies.[35] Total distance flown was 58 and 67 percent less for LT females and LT males compared to their WT counterparts. WT flies also responded differently to certain colors than LT flies.[27] WT flies had a significant preference for yellow, over orange (49-67% of yellow), while the LT flies had no significant preference (83% of yellow). WT and LT flies did not show any significant differences in their response to host detection, however.[37]

D. SEXUAL COMPETITIVENESS

Sterilized and fertile LT female flies attracted more males in an olfactometer and in cage tests under lab conditions, than did WT females. This was attributed to the larger quantity of pheromone produced by the LT females.[25] In mating tests, LT sterilized males with weights of 5.8 and 3.7 mg/fly competed equally well with their non-irradiated WT and LT mating competitors.[33]

E. BACTERIAL SYMBIOSIS AND FLORA

The bacterial complex (symbiotic or not) associated with the olive fruit fly is probably related to the insect's nutrition, by providing amino acids, vitamins or other essential nutrients.[8,38-40] Bacteria that were proposed by Petri[41] to be symbiotic were not isolated or identified by subsequent studies in WT olive flies by Yamvrias et al.[42] or Luthy et al.,[43] but the conclusion was made that one[4] or two[44] types of bacterial symbionts existed but were not identified. Tsiropoulos[45] found 8 different bacteria associated with LT flies while 16 were associated with WT ones.

V. ASPECTS OF NUTRITION AND INSECT QUALITY IN OLIVE FRUIT FLY

Heredity, environment or the interaction or both are responsible for all phenotypic variations in insects and other organisms. Nutrition is a very important environmental factor. In the past, the interaction of heredity and nutrition in causing variation was not considered to be very important, but now geneticists realize that this factor must be recognized. Olive fruit fly larvae live exclusively in olive fruits. The interaction of nutrition and other dietary factors may be very critical to their behavior and survival and should be investigated. The flies should be selected for the desired quality traits and the required nutrition should be supplied to make possible the expression of such traits. Under the existing system of olive fruit fly rearing, considerable reduction in the variation in dietary ingredients may be made by taking into consideration upper and lower limits of tolerance for certain nutrients,[11,46] preservatives[16] or other dietary factors[10,47-49] by applying standard dietary quality control procedures.[50] Consideration of the data presented here suggests that the difference between the nutritional and other dietary factors involved between the food of naturally grown and artificially reared olive fruit flies may be responsible for certain of the biological differences observed between WT and LT insects. Nutrition probably acted as an environmental factor, as a selection factor, or in combination. Unless dietary and genetic effects are distinguished and their interaction is investigated, the progress for the production of desired quality aspects for olive fruit fly for SIT will be very difficult, if not impossible. Studies in this area may be a very exciting development for SIT in the years ahead.

ACKNOWLEDGMENT

The help of Mrs H. Zografou, graduate research assistant in preparing this report, is gratefully acknowledged.

REFERENCES

1.   **Tzanakakis, M. E.,** Small-scale rearing *Dacus oleae*, in *Fruit Flies, Their Biology, Natural Enemies and Control*, Vol. 3B, Robinson, A. S. and Hooper, G., Eds., Elsevier, Amsterdam, 1989, 105.

2.   **Tsitsipis, J. A.,** Nutrition requirements, in *Fruit Flies, Their Biology, Natural Enemies and Control*, Vol. 3A, Robinson, A. S. and Hooper, G. Eds., Elsevier, Amsterdam, 1989, 103.

3.   **Tsiropoulos, G. J.,** Reproduction and survival of the adult *Dacus oleae* fed on pollens and honeydews, *Envir. Entomol.*, 6, 390, 1977.

4.   **Manoukas, A. G., Mazomenos, B., and Patrinou, M. A.,** Aminoacid content in three varieties of olive fruit, *J. Agric. Food Chemistry*, 21, 215, 1973.

5.   **Manoukas, A. G., Grimanis, A., and Mazomenos, B.,** Inorganic nutrients in natural and artificial food of *Dacus oleae* larvae (Diptera:Tephrit.), *Ann. Zool. Ecol. Anim.*, 10, 123, 1978.

6.   **Manoukas, A. G.,** Comparison of composition of the larval food of the olive fruit fly in nature and in the laboratory, in *Proceedings XXXVI International Symposium on Crop Protection*, Gent, Belgium, 643, 1984a.

7.   **Sawas-Dimopoulou, C. and Fytizas, E.,** Accumulation des acides amines libres dans le mesocarp ete l'olive apres conservation a basse temperature, *Canad. J. Biochem.*, 45, 1965, 1967.

8. **Hagen, K. S.,** Dependence of the olive fruit fly, *Dacus oleae,* on symbiosis with *Pseudomonas savastanei* for the utilization of olive, *Nature* (London), 209, 423, 1966.

9. **Scott, M. L., Nesheim, M. C., and Young, R. J.,** *Nutrition of the Chicken,* Scott, M. L. and Assoc., Ithaca, New York USA, 1969.

10. **Manoukas, A. G.,** Growth and development of the olive fruit fly larvae in yeast-free diets, in *Fruit Flies of Economic Importance,* Cavalloro, R., Ed., A. Balkema, Rotterdam, 1989a, 589.

11. **Manoukas, A. G.,** Effect of excess levels of individual amino acids upon survival, growth and pupal yield of *Dacus oleae, Z. Ang. Entomol.,* 91, 309, 1981.

12. **Manoukas, A. G.,** Effect of supplementation of larval diets upon pupal yield and adult emergence of the olive fruit fly *Dacus oleae* (Gmel.), *Z. Ang. Entomol.,* 98, 159, 1984b.

13. **Fruton, J. and Simmonds, S.,** *General Biochemistry,* 2nd Ed., John Wiley, USA, 1961.

14. **Sivapalan, P. and Gnanapragasam, N. C.,** Influence of copper on the development and adult emergence of *Homona coffearia* (Lepidoptera: Tortricidae) reared *in vitro, Entomol. Exp. & Appl.,* 28, 59, 1980.

15. **Manoukas, A. G.,** unpublished results, 1976.

16. **Manoukas, A. G. and Hasapidou, M. N.,** Tocopherol and tocotrienol composition of olives and their oil, in *Actes du Congress "Chevreul" Pour l'etude de Corps Gras,* Vol. 2, 160, 1989.

17. **Manoukas, A. G.,** unpublished data, 1990.

18. **Manoukas, A. G. and Mazomenos, B.,** Effect of antimicrobials upon eggs and larvae of *Dacus oleae* (Diptera, Tephritidae) and use of propionates as larval diet preservatives, *Ann. Zool. Anim.,* 9, 277, 1977.

19. **Bush, G. L. and Kitto, G. B.,** Research on the genetic structure of wild and laboratory strains of olive fruit fly, *F.A.O. Report. Development of Pest Management Systems for Olive Culture,* F.A.O., United Nations, Rome, 1979.

20. **Zouros, E. and Loukas, M.,** Biochemical and colonization genetics of *Dacus oleae* (Gmelin), in *Fruit Flies, Their Biology, Natural Enemies and Control,* Vol. 3B, Robinson, A. S. and Hooper, G. Eds., Elsevier, Amsterdam, 1989, 75.

21. **Economopoulos, A. P. and Loukas, M. G.,** ADH allele frequency changes in olive fruit fly shift from olives to artificial larval food and vice versa, effect of temperature, *Entomol. exp. appl.,* 40, 215, 1986.

22. **Manoukas, A. G.,** Amino acids in olive fruit fly, *Dacus oleae* (Gmelin), grown in artificial diet and in olive fruits, *J. Insect Physiol.,* 18, 683, 1972a.

23. **Manoukas, A. G.,** Free and peptide-bound aminoacids in the insect *Dacus oleae* (Gmelin) grown under natural and artificial conditions, *Comp. Biochem. Physiol.,* 43B, 787, 1972b.

24. **Vakirtzi-Lemonias, C., Karachalios, C., and Tzanakakis, M. E.,** Lipids of the adult olive fruit fly. Total lipids and major lipid fractions, *Ann. Entomol. Soc. Amer.,* 62, 1290, 1969.

25. **Haniotakis, G. E.,** Pheromone studies of the olive fruit fly, *Dacus oleae* (Gmel), in *Final report NATO, Res. Grant, 1952,* 1979, 69.

26. **Mazomenos, B. E. and Pantazi-Mazomenou, A.,** Olive fruit fly *Dacus oleae* pheromone: Comparisons of the quantity produced between a laboratory strain and a wild strain, in *Fruit Flies of Economic Importance,* Cavalloro, R., Ed., A. Balkema, Rotterdam, 1989, 151.

27. **Prokopy, R. J., Economopoulos, A. P., and McFadden, M. N.,** Attraction of wild and laboratory-cultured *Dacus oleae* flies to small rectangles of different hues, shades and tints, *Ent. Exp. App.* 18, 141, 1975.

28.   **Manoukas, S. G.,** Nuclear technology in plant protection: The case of the olive fruit fly. in *Proc. IV Bulgarian Scientific Conference on "Nuclear Methods in Agriculture and Food Industry",* Varna, Bulgaria, 20-23 Sept., 1990.

29.   **Manoukas, A. G.,** The adaptation Process and the Biological Efficiency of *Dacus oleae* larvae, in *Fruit Flies of Economic Importance,* Cavalloro, R., Ed., A. Balkema, Rotterdam, 1983a, 91.

30.   **Manoukas, A. G. and Tsiropoulos, G. J.,** Effect of density upon larval survival and pupal yield of the olive fruit fly, *Ann. Entomol. Soc. Amer,* 70, 414, 1977.

31.   **Manoukas, A. G.,** Growth and survival of *Dacus oleae* larvae under different population densities in an artificial diet, *Z. Ang. Entomol.,* 89, 259, 1980.

32.   **Economopoulos, A. P. and Tzanakakis, M. E.,** Egg yolk and olive juice as supplements to the yeast-hydrolyzate-sucrose diet for adults of *Dacus oleae, Life Sciences,* 6, 2409, 1967.

33.   **Tsiropoulos, G. J. and Manoukas, A. G.,** Adult quality of *Dacus oleae* (Gmel.) affected by larval crowding and pupal irradiation, *Ann. Entomol. Soc. Amer.,* 70, 916, 1977.

34.   **Manoukas, A. G.,** Biological and Biochemical Parameters of the Olive Fruit Fly with reference to Larval Nutritional Ecology, in *Fruit Flies of Economic Importance,* Cavalloro, R., Ed., A. Balkema, Rotterdam, 1983b, 410.

35.   **Fletcher, B. S. and Economopoulos, A. P.,** Dispersal of normal and irradiated laboratory strains and wild strains of the olive fruit fly *Dacus oleae* in an olive grove, *Entomol. exp. appl.,* 20, 183, 1976.

36.   **Zervas, G. A.,** Sexual and reproductive maturation in wild and lab cultured olive fruit flies *Dacus oleae* (Gmelin) (Diptera:Tephritidae), in *Fruit Flies of Economic Importance,* Cavalloro, R., Ed., A. Balkema, Rotterdam, 1983, 429.

37.   **Prokopy, R. J. and Haniotakis, G. E.,** Host detection by wild and lab-cultured olive flies, *Symp. Biol. Hung,* 16, 209, 1976.

38.   **Fytizas, E. and Tzanakakis, M. E.,** Some effects of streptomycin, when added to the adult food, on the adults of *Dacus oleae* (Diptera:Tephritidae) and their progeny, *Annals Entomol. Soc. Amer.,* 59, 269, 1966.

39.   **Tzanakakis, M. E. and Stavridis, A. S.,** Inhibition of development of larvae of the olive fruit fly, *Dacus oleae* (Diptera:Tephritidae), in olives treated with streptomycine, *Ent. exp. appl.* 16, 39, 1973.

40.   **Fitt, G. P. and O'Brien, R. W.,** Bacteria associated with four species of Dacus (Diptera:Tephritidae) and their role in nutrition of larvae, *Oecologia (Berlin),* 67, 447, 1985.

41.   **Petri, L.,** Untersuchung uber die Darmbakterien der Olivenfliege, *Zent. Bakt. Parasit., Infekt. und Hyg.,* 26, 357, 1910.

42.   **Yamvrias, C., Panaagopoulos, C. G., and Psallidas, P. G.,** Preliminary study of the internal bacterial flora of the olive fruit fly (*Dacus oleae,* Gmelin), *Ann. Ins. Phyt. Benaki,* 9, 201, 1970.

43.   **Luthy, P., Struder, D., Jacquet, F. and Yamvrias, C.,** Morphology in vitro cultivation of the bacterial symbiote of *Dacus oleae, Mitt. Schw. Entomol. Gesellschaft. Bull. Soc. Entomol.* Suisse, 56, 67, 1983.

44.   **Girolami, V.,** Fruit fly symbiosis and adult survival. General aspects, in *Fruit Flies of Economic Importance,* Cavalloro, R., Ed., A. Balkema, Rotterdam, 1983, 74.

45.   **Tsiropoulos, G. J.,** Microflora associated with wild and laboratory reared olive fruit flies, *Dacus oleae* (Gmel.), *Z. Ang. Entom.,* 96, 337, 1983.

46.   **Manoukas. A. G.,** Effect of excess levels of inorganic salts upon survival, growth and pupal yield of *Dacus oleae* (Gmel.) larvae, *Z. Ang. Entom.,* 93, 208, 1982.

47.   **Manoukas, A. G.,** Biological characteristics of *Dacus oleae* larvae (Diptera, Tephritidae) reared in a basal diet with variable levels of ingredients, *Ann. Zool. Ecol. Anim.* 9, 141, 1977.

48.   **Manoukas, A. G.,** The effect of water content of the growth media upon the performance and nutritional ecology of the olive fruit fly larvae, Amsterdam, The Netherlands, Proceedings, 1986, 283.

49.   **Manoukas, A. G.,** Amino acid mixtures for replacing soy hydrolysate in larval diets of *Dacus oleae* (Gmel.) (Dipt., Tephritidae), *J. Appl. Entomol.*, 108, 102, 1989b.

50.   **Manoukas, A. G.,** Dietary control in insects. The case of the olive fruit fly, *Dacus oleae* (Gmel.), in *Proc. Workshop of Int. Organ. Biol. Control, Global group on "Quality Control of Mass Reared Arthropods",* Wageningen, NL, March 25-28, 1991, 174.

Chapter 7

ADVANCES IN MEASURING QUALITY AND ASSURING GOOD FIELD PERFORMANCE IN MASS REARED FRUIT FLIES

C. O. Calkins, K. Bloem, S. Bloem,
and D. L. Chambers

I. INTRODUCTION

Several species of tephritid fruit flies are major threats to many agricultural fruit crops. More importantly, they could become serious quarantine pests for countries that do not have endemic populations of the species in question. To prevent introductions of these pests, quarantine laws require that all host fruits from exporting countries harboring certain species of fruit flies be treated with fumigation or some other quarantine treatment. Another option that is just beginning to be accepted is the maintenance of a zone in the citrus growing area that is kept free of fruit flies, at least during the picking season.

New infestations, in countries that are free of these fruit flies, are usually attacked vigorously by insecticidal bait sprays and/or by the sterile insect technique (SIT) to eradicate the flies as soon as possible. Establishment of a new species in a fly-free country or zone jeopardizes the export of all host fruits from those areas if immediate attention is not given to eradicating the flies.

A deterrent to the widespread use (acceptance) of insecticidal bait sprays by growers is the widespread concern for the use of insecticides in the environment. This has placed extreme pressure for eradication solely on the use of SIT to eradicate local and widespread populations of fruit flies. This technique is quite effective only if the sterile males are successful in interacting and mating with the target wild females. To assure that sterile males are competitive with wild males requires monitoring the quality of the fly. This is the responsibility of a quality control unit now present in most fruit fly production factories.

Because quality control is so important in mass production of sterile flies, a definition of quality control must be given. Quality Control (QC) is a system for maintaining desired standards in production and in a product. Without feedback for procedures and defects, improvements in quality cannot be accomplished. This closed loop feed-back system provides information required to modify and improve or develop processes and procedures that ultimately produce more acceptable and reliable products. In an insect mass rearing program, we are indeed producing a product (the sterile male) that is expected to have certain characteristics that allow it to fulfill a specific intended purpose. In most cases, this purpose is for sterile released insects to interact with a wild population to bring about suppression or eradication.

The major concern in using SIT is that the technique may begin to fail without the realization of program personnel until it is too late to do anything about it. Laboratory bioassays monitor the output of mass production to assure that a stable product is always being produced. However, they cannot fully insure that the flies produced will perform as expected in the field. The bottom line in an SIT program is that sterile flies interact with the wild population effectively in the manner anticipated. The types of tests needed to ascertain this interaction must include wild flies and are, therefore, not easy to conduct. Access to large numbers of wild flies is necessary to measure such interactions with accuracy and confidence.

Development of a quality control system for mass reared fruit flies was reached in 1978 when a laboratory QC program was initiated by Boller et al.[1] for inclusion in the eradication program for Mediterranean fruit fly in Mexico. Funding for the program was furnished by the International Atomic Energy Agency (IAEA). Various field aspects of quality control associated with the laboratory bioassays were evaluated in a coffee finca in Guatemala in November, 1978.[2] They were refined further in a training course in Castellon, Spain in 1979.[3] Since that time, advances in fruit fly quality control are achieved through a series of workshops sponsored by the International Organization of Biological Control (IOBC) Global Working Group.

In 1984, a three-year pilot test sponsored by USDA, Agriculture Research Service, was initiated in Guatemala to implement technical and managerial systems for quality control in a medfly sterile release program.[4] Specifications and tolerances for process and product control were introduced into the QC system for the first time. This resulted in the development of a QC manual by and for USDA-APHIS for use in producing or purchasing medflies for sterile male release programs.[5]

A remaining weakness in fruit fly quality control is the lack of a rapid and uncomplicated method of evaluating the effects of sterile fruit fly releases on the indigenous population. This paper reviews the progress that has been made in attempting to develop bioassays that evaluate the effectiveness of mass reared flies in the field. It also points out the problems that could plague SIT programs if constant monitoring of the interactions between sterile and wild flies does not occur.

II. LABORATORY BIOASSAYS

A. MATING COMPETITIVENESS
1. Refractory Periods

Because medfly females mate very soon after emergence, virgin wild females make up only a small portion of the total population present in the field. Therefore, sterile male flies released for the purpose of mating with these virgin females have only a small window of opportunity. Once young females are mated, they probably do not mate again until they have used most of the sperm from the first mating. Only a small percentage of the females probably survive long enough to exhaust their sperm because of the many mortality factors in the field.

The identification of factors influencing the length of time a female is refractory is important in determining the management of the SIT program. Two laboratory studies were conducted in Guatemala that examined: (1) the effect of a male's reproductive status during a female's first mating on her future mating receptivity,[8] and (2) the influence of male size during initial matings on the length of a female's refractory period.[9]

Results from the first study showed that remating of females mated initially with sterile males (whether virgin or once-mated) was much greater than among females that had mated with fertile males. For example, an average of 64.5% and 58.4% of the females initially mated with virgin sterile and once mated sterile males remated, compared with only 36.8% of the females originally mated with VF males remating. This would suggest that fertile males are inherently of a higher quality than sterile males relative to producing a longer refractory period in females. However, results from the second study indicated that the inadequacy of sterile males to produce a long refractory period could at least be partially compensated for by increasing male size. Fewer females remated when the initial mating was with a large virgin fertile male (51% remating) than when it was with a small virgin fertile male (70% remating). Likewise, of those females that did remate, the subsequent refractory period was longer if mated with large males.

Multiple matings are of concern for SIT programs because they increase the chance that a gravid female will encounter and mate with a fertile wild male.[6] This may enable the female to lay fertile eggs even if she initially mated with a sterile male. Saul et al.[7] showed that offspring from multiply-mated females may not be the result of the initial mating, but could result from subsequent matings because of sperm mixing and/or sperm competition.

Because a long postmating refractory period is beneficial to the functioning of the SIT, the tests described above continue to support a growing body of literature that recommends the production of large flies at mass rearing facilities as a means of assuring high quality. In addition, it has already been established that lower doses of radiation result in higher quality flies.[10, 11] If males irradiated at lower doses also produce a longer refractory period in wild females than do males given the higher recommended dose, it may become a management issue in areas like Central America where there is a large established population. Then it must be determined whether it is better to produce males with no residual fertility but which result in matings in which the females remate after a short period of time, or allow some degree of fertility (e.g. 2-3%) but have higher quality males that produce longer female refractory periods. Hooper and Katiyar[12] determined that as irradiation doses increased, male competitiveness decreased. Those females in competitive mating tests with males irradiated at 3-, 5-, 7-, 9-, 11-krad and with fertile males had no significant difference in egg hatch even though the males with the lower irradiation doses had higher residual fertility. When the irradiation dose increased above 13-krad, a greater percentage of eggs hatched.

III. FIELD TESTS

A.   DISPERSAL RATE

Good dispersal ability by mass reared flies is critical to the success of an SIT program. Poor fliers will be lost quickly in hostile environments or will be unable to locate and compete with wild populations at mating sites. In an effort to develop a field component complementary to the more clinical laboratory assessment of quality parameters published by Boller et al.,[1] a series of biological assays of behavioral competence were outlined by Chambers et al.[2] Among these, release-recapture tests were presented more as general procedures and not as standardized protocols. Therefore, based on recommendations in Chambers et al.[2] and recognized by the International Atomic Energy Agency, more detailed procedures were developed for routine field assessment of the dispersal and survival ability of mass reared Mediterranean fruit flies produced in Guatemala. A number of experiments that demonstrate the type of information that can be gained from the use of these more detailed protocols are also presented.

1.   Techniques in Coffee Fincas in Guatemala

A 200 square meter plot of mature coffee plants was set up for the dispersal study. Within the plot, an 8x8 grid of trimedlure-baited Jackson traps was emplaced, with traps spaced at 25 meter intervals with a 12.5 meter border left around the plot. Traps were hung at a height of about 1.5 meters in coffee plants.

Two distinctly different strategies for releasing flies were used, depending upon the nature of the study. For dispersal tests, traps were set out in mid-afternoon followed by the release of medflies in late afternoon (ca. 5-6 P.M.) from one point in the center of the plot. At least ten thousand medflies of each group (e.g. production lots, strains, etc.) being compared were released. For longevity and other release-recapture experiments, sixteen release sites were used in a 4x4 grid with 50 meter intervals.[13] Thirty released males/Ha (=120 males /plot or 7-8 males/release site) were found to be very close to the detection limit for this experimental design, while 3,000 males/Ha (=12,000/plot or 750/release site) resulted in 90-100% of the traps capturing at least one medfly.

Typical recapture rates for both release methods were between 2 and 4 percent. Two to three replicates of these tests gave consistent, comparable results. Trends for a given comparison were consistent over time, but absolute values varied widely and could not be compared directly from one time to another. We conducted standard laboratory flight ability and longevity tests on the same lots of flies to maintain relevancy on tests conducted on different dates and to correlate findings with laboratory bioassays.

2. Dispersal Ability of Medflies Produced From Recycled Diet

Recycling of a percentage of the spent medfly diet was studied in an effort to reduce waste and save expenses at the Guatemala mass rearing facility. The recycling technique consisted of combining a larval starter diet[14] with spent, heat-sterilized larval rearing media that had 33% fresh ingredients added.

The flies produced using recycling technology were then examined to ascertain any potential effects on fly quality. Both laboratory and field tests indicated that flies produced on recycled diet performed equal to or better than flies produced on the fresh diet. In laboratory quality control tests, average pupal weight, and percent eclosion were similar for both groups, while percent flight ability was 79.3 for flies from the recycled diet and 68.0 for flies from normal diet. In the field test, twice as many medflies from recycled diet were recaptured than flies from normal diet and they were found in traps over a greater period of time than flies produced from normal diet.

3. Competitiveness and Survival of Flies of Different Sizes

A growing body of literature consistently shows that larger flies out-perform smaller flies in laboratory tests.[15,16] However, the question of whether field performance is really affected continues to be discussed because this has important implications in terms of mass-rearing costs. Competitiveness seems to be related to mating success, another measure of quality. One behavior in which size is particularly important is in the competition for hierarchical sites in the lek. Large males seem to win more bouts for territory with other males than do small ones. The fact that flies must compete for desirable sites in the lek makes this attribute very important in an SIT program. Large males also are preferred by females in mating choices.[17,18] This attribute may be related to the amount of material of a nutrient nature transferred from the male accessory glands to the female during mating. Alternatively, the female may use this as an indication of male fitness.

To evaluate survival of smaller versus larger mass reared flies in the field, release/recapture tests were conducted in Guatemala utilizing flies from different pupal weight classes.[19] In each of three separate releases, a higher percentage of large flies (8.0-8.5 mg pupae) were recaptured than either medium sized (6.5-7.0 mg pupae) or small flies (5.0-5.5 mg pupae), with large flies also tending to be caught over a longer period of time than small flies. Concurrent laboratory QC tests also found that larger flies tended to have higher emergence rates and flight ability indices, and were significantly longer lived than smaller flies when deprived of food and water.[17]

4.   Distribution of Wild and Laboratory Flies in the Field

Properly designed release-recapture tests not only provide information on dispersal ability and longevity of released flies, but may reveal the ability of these flies to orient to host plants[2] or to areas that provide food or protection. In this study, Jackson traps were used to identify three areas within a coffee finca that consistently had different population densities of wild medflies. Release-recapture experiments were then conducted in these same areas (using the 16 release sites/plot scenario) to see if relative levels of sterile fly recaptures corresponded to the levels of wild fly catches. The trapping survey prior to the releases indicated that wild population levels were highest in one identified area (Plot A) and lowest in another (Plot C). The sterile flies were released uniformly throughout the plots. The numbers of sterile flies captured during that first week in each area following the release showed a similar trend. For example, of the wild flies, 2,611 (61%) were caught in Plot A; 1,407 (33%) in Plot B and 254 (6%) in Plot C. Of the sterile flies, 312 (48%), 213 (33%), and 121 (19%) were captured in Plots A, B, and C, respectively. The higher recapture of sterile flies in Plot C was due probably to recapture before the flies had a chance to redistribute themselves adequately. This trend also held for captures within a given plot. Those sections where higher or lower numbers of wild flies were consistently caught also yielded higher and lower numbers of released sterile flies trapped. Again, for example, in Plot A, about 8% of the wild and 3% of the sterile flies were caught in each of rows 1 to 6, 12% of the wild and 16% of the sterile flies were caught in row 7, and 40% of the wild and 66% of the sterile flies were caught in Row 8. These results suggest that mass-reared flies from the Guatemala facility have the ability to disperse to and discern suitable habitats, and that their habitat preferences are similar to those of wild medflies. It is important to the SIT program that the sterile released flies inhabit the same territory as the wild flies. Thus, this test is a valuable addition to the array of field tests that compare the behavior of laboratory flies with that of wild flies.

5.   Recapture Rates of Male-only Releases

The underlying assumption in the SIT is that matings by sterile males with wild females is the principal mode of action. Sterile females are usually considered inconsequential or deleterious as they may produce ovipositional stings in fruit.[20] Therefore, much effort has been invested in the development of genetic sexing or "male-only" strains. In the following experiment, the influence of females on sterile fly dispersal was examined by comparing recapture data from male-only releases with that of male/female releases.   Releases were made in plots of coffee plants in Guatemala during the rainy season, when wild medfly populations were extremely low.   Average recapture rates of sterile males from three releases of males only and three using both males and females were 2.5% and 4.5%, respectively.   In addition to the difference in recapture rates when only males are used, 99% of the flies were caught during the first week and no flies were caught after two weeks. Where males and females were released together, males continued to be caught into the 5th week. It appeared that males in the absence of females tended to disperse away from the release site, presumably in search of females. During the dry season, when wild medfly populations were high, obvious differences between recapture rates were not observed between the two types of releases.

6.   Evaluation of the Effect of Sterile Releases on the SIT Program

Finally, the bottom line for determining the effectiveness of the SIT is the continuous suppression or reduction of the wild population. In an area-wide suppression or eradication program, the wild population is usually reduced by a bait spray, after which sterile flies are released. The first indication that the program is not working is when the wild population continues to increase. By the time this is evident, the program may be in dire straights.   An earlier indication that the program may not be functioning properly can be discovered in one of two ways. If two or more species of fruit flies occur in the same area, the population levels of one designated as the target species can be compared with the other species (control) by monitoring trap catches before and after the bait spray and after the sterile releases. One would expect the numbers of the control species to rebound at some time after the bait spray providing no other generalized control is applied. The number of the specimens of the target species trapped should not increase. In situations where only one species (target species) occurs, the same information can be acquired by applying the bait spray over a larger area than that where the sterile flies will be released, then monitoring the populations in the two areas before and after treatments. Again, one should expect the population, in the area sprayed with no sterile releases, to rebound while the population in the sprayed area undergoing sterile releases to remain suppressed. There will be a brief period of population increase in the target area immediately after the bait spray because of the emergence of new adults that were in the larval and pupal stage at the time of the spray. This fact should be taken into account.

By using a control species or a control area to evaluate the sterile releases, this identifies the possibility that some other factor may be suppressing the target population. This would not be apparent without such controls and the possibility that ineffective flies were being released would not be evident until the effect of the suppressing factor(s) disappear.

IV. MATING COMPATIBILITY

Mating incompatibility between wild and sterile released fruit flies may be one of the major causes of failure of SIT programs. Unfortunately, it is also one of the most difficult factors to recognize. Even though resistance to insecticides is a well documented occurrence, development of resistance to the sterile insect technique is less well understood and should be investigated. Mating incompatibility between sterile flies and wild flies could conceivably occur if certain conditions exist. For example, a prolonged exposure of the sterile flies to wild populations might lead to selection against interstrain matings. This situation could occur if eradication programs take too long to achieve their goal, usually through the use of ineffective sterile flies (i.e. low quality), inadequate calculations of overflooding ratios or through the continued use of sterile flies as a "biological barrier" between a fly-free zone and an infested area. Incompatibility of the mass reared population might also occur through genetic drift where pheromone composition, acoustical signals and/or visual cues are no longer capable of bringing the sexes together.

Resistance could also occur through changes in mating periodicity. Most medflies mate in the morning when light intensity and temperature reach a certain threshold. Almost all matings are completed by noon, although in some wild populations, mating may extend into the early afternoon. In some laboratory strains, the matings rarely extend past noon. If an eradication program is allowed to continue for an extended period of time, there could be enough selection pressure against the wild population that the only successful fertile matings would shift from forenoon to afternoon, thereby effectively eliminating the possibility of interstrain matings. There is evidence that such a temporal change in mating times is occurring in the laboratory colony of the Caribbean fruit fly, *Anastrepha suspensa* (Loew) in Florida.[21] Wild populations of this species normally mate in the late afternoon near dusk. The laboratory strain has been reared under continuous bright light for several generations without a dusk period programmed into the lighting system. As a consequence, laboratory reared flies now mate faster in bright light as opposed to normal testing conditions of dim light. Preliminary field cage trials with both laboratory and wild flies indicate that the laboratory strain begins to mate about 2 P.M., while the wild flies do not start mating until several hours later. Thus, the mating time of laboratory flies is asynchronous with that of the wild flies.

Other tests have been developed to evaluate the compatibility of mass reared and wild flies, based primarily on the mating behavior of the wild population. Some of the specific behavioral traits examined were as follows:

A.   ACOUSTIC SIGNALS

A comparison was conducted for the relative ability of acoustic signals produced by sterile and wild males to attract wild females by Sivinski et al.[22] They discovered that a significant difference existed in the fundamental frequency of calling songs between wild males and fertile laboratory reared males where laboratory males had a lower frequency (possibly associated with their larger size). Interestingly enough, there was no difference in the fundamental frequency of acoustic signals between wild males and sterile laboratory males.

B.   PHEROMONE ATTRACTION

Response to pheromone produced by calling males is another way of testing for interstrain incompatibility. A test was developed to compare attractiveness between sterile laboratory reared medfly males and wild males. The flies were confined separately in small cages suspended inside a large field cage[23] with empty cages serving as controls. Wild females were then released into the large cage and allowed to distribute themselves at random throughout the cage. At ten minute intervals, the number of females that had landed on the test cages were counted and gently blown off using an aspirator. The test lasted two hours. If females were attracted in equal or greater numbers to test cages containing sterile flies, the sterile fly population was considered to be pheromone-compatible. Results showed that sterile males attracted more wild females than did the wild males.[24] More recently, Heath and Manukian[25] have developed techniques to measure the amount and ratio of pheromone components produced by sterile and wild flies. This technique has been tested on both medflies and Caribbean fruit flies.

C.   GENETIC ANALYSIS

Where small populations of wild flies are present, as in the case of new infestations in the United States, compatibility tests between such populations and the sterile strains used to control them become quite difficult. In this case, it becomes important to establish the origin of the infesting population. In the past, the detection of unique rare alleles, through isozyme techniques, could be used as genetic markers to determine the origin of populations.[26] The advent of mitochondrial and nuclear DNA analyses have allowed for more precise identification of populations with the use of smaller samples.[27,28] Medfly introductions in the United States have the greatest potential of originating in Hawaii and Latin America. Therefore, mass reared medfly colonies that support these emergency SIT programs in the continental U.S. (such as those in Hawaii, Mexico, and Guatemala) should be tested annually for compatibility with wild populations from these two geographical areas.

In summary, to prevent development of resistance to SIT, constant vigilance through monitoring the compatibility of laboratory and wild flies must be maintained. If genetic drift proceeds too far, then periodic renewals of colonies from wild target populations may be necessary.

V. CONCLUSIONS

Results presented herein strongly suggest that the production of large flies is a desirable goal in a mass rearing facility. It is a relatively easy adjustment to make in a rearing factory, for example, by slightly reducing the ratio of eggs to diet. Also, there are nutritional and temperature aspects that influence size that should be considered in the rearing process. However, when production is proceeding normally, reducing egg seeding rate will have a dramatic effect. It must be impressed upon rearing personnel that the quality of the flies far outweighs the numbers produced.[29]

Quality control is a dynamic process. Improvements in existing tests and development of new ones should be uppermost in everyone's mind. The process should additionally focus on new areas that could yield greater insight into the effectiveness of the flies produced, such as monitoring of the success of sterile matings among wild females.

Quality control is one of the most useful management tools available to the SIT program manager. It serves as his "eyes and ears" and allows prediction and adjustment of strategies to gain maximum effectiveness in the rearing program. It also permits continuous monitoring of the rearing process with a minimum of effort.

ACKNOWLEDGEMENT

Some of the work reported here was funded by the International Atomic Energy Agency, Vienna, Austria through a grant to Guatemala, supervised by Dr. D. L. Chambers (Project No. 518/GUA/5/007/404F). We would like to acknowledge the assistance and participation in Guatemala of N. Rizzo and E. Muniz in the planning, conduct and analysis of the experiments described herein. We also thank Steve Ferkovitch, Patrick Greany, and Tim Holler for valuable suggestions to the manuscript.

REFERENCES

1. **Boller, E. F., Katsoyanous, B. I., Remund, U., and Chambers, D. L.,** Measuring, monitoring and improving the quality of mass-reared Mediterranean fruit flies, *Ceratitis capitata* (Wied.) 1. The RAPID quality control system for early warning, *Z. Ang. Entomol.*, 92, 67, 1981.

2. **Chambers, D. L., Calkins, C. O., Boller, E. F., Itô, Y., and Cunningham, R. T.,** Measuring, monitoring and improving the quality of mass-reared Mediterranean fruit flies, *Ceratitis capitata* (Wied.). 2. Field tests confirming and extending laboratory results, *Z. Ang. Entomol.*, 95. 285, 1983.

3. **Calkins, C. O., Boller, E. F., Chambers, D. L., and Itô, Y.,** Quality control in *Ceratitis capitata*: Training manual for the International Course on Quality Control in *Ceratitis capitata*, held in Castellon, Spain, September 17-27, 1979, 1980.

4. **Calkins, C. O., Ashley, T. R., and Chambers, D. L.,** Implementation of Technical and Managerial Systems for Quality control in Mediterranean fruit fly (*Ceratitis capitata* (Wied.)) Sterile Release Programs, *Bull. Entomol. Soc. Amer.* (In preparation).

5. **Brazzel, J. R., Calkins, C. O., Chambers, D. L.,and Gates, D. B.,** Required quality control tests, quality specifications, and shipping procedures for laboratory produced Mediterranean fruit flies for sterile insect control programs, USDA-APHIS-PPQ Report 81-51, 1986, 27 pp.

6. **Nakagawa, S., Farias, G. J., Suda, D., Cunningham, R. T., and Chambers, D. L.,** Reproduction of the Mediterranean fruit fly: Frequency of mating in the laboratory, *Ann. Entomol. Soc. Amer.*, 64, 949, 1971.

7. **Saul, S. H., Tam, S. Y. T., and McInnis, D. O.,** Relationship between sperm competition and copulation duration in the Mediterranean fruit fly (Diptera: Tephritidae), *Ann. Entomol. Soc. Amer.*, 81, 498, 1988.

8. **Bloem, K. A., Bloem, S., Rizzo N., and Chambers, D. L.,** Female refractory period: Effect of male reproductive status, in *Fruit Flies: Biology and Management,* Aluja, M. and Liedo, P., Eds., Springer-Verlag, New York,1993, 189.

9. **Bloem, S., Bloem, K. A., Rizzo N., and Chambers, D. L.,** Female refractory period: Effect of first mating with sterile males of different sizes, in *Fruit Flies: Biology and Management,* Aluja, M. and Liedo, P., Eds., Springer-Verlag, New York,1993, 191.

10. **Hooper, G. H. S.,** Sterilization of the Mediterranean fruit fly with gamma radiation: Effect on male competitiveness and change in fertility of females alternately mated with irradiated and untreated males, *J. Econ. Entomol.*, 65, 1, 1972.

11.   **Calkins, C. O., Draz, K. A. A., and Smittle, B. J.**, Irradiation/sterilization techniques for *Anastrepha suspensa* Loew and their impact on behavioural quality, in *Proc. Intl. Symp. Modern Insect Control: Nuclear Techniques and Biotechnology*, Lindquist, D. L., Ed., Vienna, Austria, 1988, 299.

12.   **Hooper, G. H. S. and Katiyar, K. P.**, Competitiveness of gamma-sterilized males of the Mediterranean fruit fly, *J. Econ. Entomol.*, 64, 1068, 1971.

13.   **Bloem, K. A., Bloem, S., Rizzo N., and Chambers, D. L.**, Field assessment of quality: Release-recapture of mass-reared Mediterranean fruit flies (Diptera: Tephritidae) of different sizes. Environ. Entomol. (submitted).

14.   **Fay, H. A. C.**, A starter diet for mass-rearing larvae of the Mediterranean fruit fly *Ceratitis capitata* (Wied.), *J. Appl. Entomol.*, 105, 496, 1988.

15.   **Churchill-Stanland, C., Stanland, R., Wong, T. T. Y., Tanaka, N., McInnis, D. O., and Dowell, R. V.**, Size as a factor in the mating propensity of Mediterranean fruit flies, *Ceratitis capitata* (Diptera: Tephritidae), in the laboratory, *J. Econ. Entomol.*, 79, 614, 1986.

16.   **Krainacker, D. A., Carey, J. R., and Vargas, R. I.**, Size-specific survival and fecundity for laboratory strains of two tephritid (Diptera: Tephritidae) species: Implications for mass-rearing, *J. Econ. Entomol.*, 69, 136, 1989.

17.   **Burk, T. and Webb, J. C.**, Effect of male size on calling propensity, song parameters, and mating success in Caribbean fruit flies, *Anastrepha suspensa* (Loew) (Diptera: Tephritidae), *Ann. Entomol. Soc. Amer.*, 76, 678, 1983.

18.   **Webb, J. C., Sivinski, J., and Litzkow, C.**, Acoustical behavior and sexual success in the Caribbean fruit fly, *Anastrepha suspensa* (Loew), *Environ. Entomol.*, 13, 650, 1984.

19.   **Bloem, K. A., Bloem, S., and Chambers, D. L.**, Field assessment of quality: Release-recapture of mass-reared Mediterranean fruit flies (Diptera: Tephritidae) of different sizes, *Environ. Entomol.*, 1993 (Submitted).

20.   **McDonald, P. T. and McInnis, D. O.** *Ceratitis capitata* (Diptera: Tephritidae): Oviposition behavior and fruit punctures by irradiated and untreated females in laboratory and field cages, *J. Econ. Entomol.*, 78, 790, 1985

21.   **Calkins, C. O.**, unpublished data.

22.   **Sivinski, J., Calkins, C. O., and Webb, J. C.**, Comparisons of acoustic courtship signals in wild and laboratory reared Mediterranean fruit fly *Ceratitis capitata* (Wied.), *Fla. Entomol.*, 72, 212, 1989.

23.   **Calkins, C. O. and Webb, J. C.**, A cage and support framework for behavioral tests of fruit flies in the field, *Fla. Entomol.*, 66, 512, 1983.

24.   **Calkins, C. O.**, unpublished data.

25.   **Heath, R. R. and Manukian, A.**, Development and evaluation of systems to collect volatile semiochemicals from insects and plants using a charcoal-infused medium for air purification, *J. Chem. Ecol.*, 18, 1209, 1992.

26.   **Kourti, A., Loukas, M., and Economopolous, A. P.**, Population genetics of the Mediterranean fruit fly *Ceratitis capitata* (Wied.), in *Genetic Sexing of the Mediterranean Fruit Fly*, IAEA Panel Proceedings Services, Vienna, 1990, 7.

27.   **Sheppard, W. S., Steck, G. J., and McPheron, B. A.**, Geographic populations of the medfly may be differentiated by mitochondria DNA variation, *Experientia*, 48, 1010, 1992.

28.   **Steck, G. J. and Sheppard, W. S.**, Mitochondrial DNA varia/tion in *Anastrepha fraterculus*, in *Fruit Flies Biology and Management*, Aluja, M. and Liedo, P., Eds., Springer-Verlag, New York,1993, 9.

29.   **Calkins, C. O. and Ashley, T. R.**, The impact of poor quality of mass-reared Mediterranean fruit flies on the sterile insect technique used for eradication, *J. Appl. Entomol.*, 108, 401, 1989.

Chapter 8

MUTANTS, CHROMOSOMES, AND GENETIC MAPS IN THE MEDITERRANEAN FRUIT FLY

Y. Rossler, A. Malacrida, and M. C. Zapater

I. INTRODUCTION

The first morphological mutant in the medfly (Mediterranean fruit fly, *Ceratitis capitata*, Diptera: Tephritidae) was reported from Hawaii in 1973.[1] It was an autosomal recessive white-eye. Simultaneously, genetic studies of the medfly were carried out in Italy at the Genetics Inst., Univ. of Bologna where an autosomal recessive orbital bristle mutant was isolated[2] and a pupal color variant (dark pupa) was found governed by an autosomal dominant allele, and modified by food and environmental factors.[3] Systematic search for mutants and directed genetic research of the medfly were initiated in the late 1970s, by the Insect & Pest Control Section of the FAO/IAEA Joint Division, in a research program on the Sterile-Insect-Technique. These studies, which were initiated by entomologists with some interest in genetics, involved later geneticists, cytogeneticists, and recently also molecular biologists. Since the isolation of the first few mutants in the early 1970's, 85 morphological mutants and 33 biochemical and functional loci have been isolated, described, and studied.[4-10]

This chapter summarizes the present status of the formal genetics of the medfly, updates the information on the morphological, biochemical, and functional loci, that have been isolated and studied in the various laboratories, presents the genetic maps known today, and points at the needs for further studies.

II. THE CHROMOSOMES

Mendes, in 1958, described the mitotic chromosomes of the medfly.[11,12] There are 5 pairs of autosomes and a pair of sex-chromosomes (the male being heterogametic).[12-18] Radu et al.[12] numbered the autosomes from 2 to 6, based on the descending relative length. The sex chromosomes was marked as chromosome number 1. These definitions were accepted by all scientists.

Polytene chromosome studies were initiated by Bedo[13,14] who investigated the trichogen cells of the male's superior orbital bristle (SO bristle), and by Zacharopoulou[17,18] who studied the salivary gland chromosomes of the third larval instar. Both scientists presented excellent descriptions of the polytene chromosomes. In 1989 these data were correlated to the mitotic chromosomes, and the chromosomes marked by genetic markers.

0-8493-4854-4/94/$0.00+$.50

The Adh locus was the first functional locus to be mapped on a given chromosome. An Adh-Y translocation line, originally constructed for genetic sexing in SIT projects, was studied by Zapater and Robinson.[19] They mapped the Adh locus on the second chromosome (the largest autosome). Subsequent studies recognized this Adh locus as Adh1.[20] Busch-Petersen and Southern[21] used the ap-dc loci (apricot eye, double chaetae) in recombination suppression lines induced by irradiation and mapped them on the fourth chromosome. Zacharopoulou[18] made a major breakthrough through studies of Y-autosome translocation lines such as 'line 127' (Y-dark pupa), and 'line wp23' (Y-white pupa). The dp locus was mapped on the third chromosome, and the wp locus - on the fifth chromosome. Thus, 4 of the 5 autosomes were now marked by morphological mutants. An additional and more important step was the correlation of the mitotic and the polytene chromosomes, so that a clear and complete picture of the chromosomes of the medfly is now available.

The road to further progress was now cleared, and many of the available morphological, biochemical, and functional loci could be mapped on the now described chromosomes.

III. MORPHOLOGICAL MUTANTS

As stated previously, a total of 85 morphological mutants were reported since 1973. Thirty-seven loci were mapped on chromosomes (Table 1), the data of 17 were published in refereed scientific publications.[7,10,22-29] The data of 22 mutants have not been published yet, but their inheritance has been relatively well studied and publication is pending. Of the remaining 46 mutants which have not been mapped yet (Table 2), the data of 6 mutants has been published.[1-3,30-32]

Forty mutants have been either reported only at research coordination meetings with no known follow-up of the authors, or are still under study in the respective laboratories.

At least three markers were reported by more than one laboratory and might actually be identical. These are the body color markers ye[26] and bwb,[33] the eye color mutants rl,[30] rb,[26] and maroon,[34] and the eye color mutants white-eye,[34] we,[26] and white-eye.[1]

Thirty-three of the reported mutants involve eye color, 11 - eye shape, 11 - body color patterns and shape, 11 - pupal color and 4 shape, 9 - hair and setae, 8 - wing mutants and 2 - antennae.

IV. BIOCHEMICAL AND FUNCTIONAL LOCI

At present, 33 functional gene markers have been assigned to five of the six linkage groups which correspond to the autosomes of the medfly. The correlation between morphological linkage groups[10] and biochemical and

Table 1
Morphological Loci Mapped in *Ceratitis capitata*

Symbol	Description	Mode of inherit.	Chrom-osome
r	red eye color[28]	rec.	X
Oe	oval eye[35]	Dom.	2
"ye"	yellow body[36]	rec.	2
	reduced eye[34]	rec.	2
	dark antenna[34]	rec.	2
ry	rosy eye color[27]	rec.	2
ew	eroded wings[25]	rec.	3
dp	dark pupa[37]	rec.	3
stubby	antennae 35-50%[34]	rec.	3
dc	double SO bristle[37]	rec.	4
B	Bar eye[24]	Dom.	4
sb	stout bristles[35]	rec.	4
ra	rosa eye color[38]	rec.	4
vg	vestigial wings[35]	rec.	4
sk	sparkling eye[10]	rec.	4
Sp	spotty abdomen[25]	Dom.	4
Sr	Sargeant abdomen[25]	Dom.	4
db	dark-brown eye color[35]	rec.	4
ap	apricot eye color[37]	rec.	4
lt	light-red eye color[28]	rec.	4
wp	white pupa[22]	rec.	5
flat	flat eye facets[34]	rec.	5
ro	rough eye[26]	rec.	5
h	harpoon SO bristle[35]	rec.	5
Cy	curly wings[35]	Dom.	5
by	burgundy eye color[39]	rec.	5
v wing	wing decreased[35]	rec.	5
or	orange-red eye color[40]	rec.	5
ye	yellow body[26]	rec.	5
bwb	brown body[33]	rec.	5
rb	ruby eye color[26]	rec.	5
we	white eye color[26]	rec.	5
w	white eye color[33]	rec.	5
Pr	dark-purple eye color[28]	Dom.	5
gr	garnet eye color[24]	rec.	6
fs	fragile setae[39]	rec.	6
bo	brown-orange eye color[24]	rec.	6
ye2	yellow eye color[33]	rec.	6

Table 2
Morphological Loci Not Mapped Yet in *Ceratitis capitata*

Symbol	Description	Mode of Inher.
da	dark abdomen I[41]	auto, rec.
di	dark imago II[41]	auto, rec.
tg	thoracic groove[35]	auto, rec.
ra	rose abdomen[41]	auto, rec.
nig	dark imago I[41]	auto, rec.
dd	dark abdomen III[41]	auto, rec.
mb	mahogany red eye color III[41]	
pl	plum eye color[41]	auto, rec.
mr	mahogany red eye color[41]	auto, rec.
sc	scarlet eye color[41]	auto, rec.
si	brilliant red eye color III[41]	
ub	ultramarine eye color[41]	auto, rec.
	white eye color[34]	auto, rec.
	flat red green tinge eye color[34]	auto, rec.
maroon	dull red eye color[34]	auto, rec.
ru	rubber eye color[30]	auto, rec.
ro	rose eye color[42]	auto, rec.
	white eye color[1]	auto, rec.
ca	carmine eye color[41]	auto, rec.
rl	reflectionless eye[31]	auto, rec.
bi	brilliant red eye color I[41]	
vr	red vinacaous eye color[41]	auto, rec.
vi	brilliant red eye color II[41]	
sw	sweet cherry eye color[41]	auto, rec.
	anomalous ommatidia[41]	auto, rec.
pr	pearled eye shape[41]	auto, rec.
ey	eyeless[35]	?
Pe	popeye eye shape[35]	auto, Dom.
bl	bilobated eye[41]	auto, rec.
nig	black pupa I[41]	auto, ID
lp1	long pupa[41]	auto, rec.
Dp	dark pupa[3]	auto, Dom.
lp2	long pupa[41]	auto, rec.
b	bended pupa[41]	auto, rec.
bs	black pupa IV[41]	auto, ID
bp	black pupa II[41]	auto, ID
bn	black pupa III[41]	auto, ID
"hl"	hazel (brown dark) pupa[38]	auto, rec.
ob	club-like seatae[2]	auto, rec.
C	chaetaless[41]	auto, ID
hl	hairless[35]	auto, rec.
el	elbowed seatae[35]	auto, rec.
s	supernumerary seatae[32]	auto, rec.
break	intercostal wing vein broken[34]	auto, rec.
nw	notched wings[35]	auto, rec.
"cy"	curly wings[36]	auto, rec.
fw	folded wings[41]	auto, rec.

functional loci has already been assessed.[43] Each of the linkage groups carries structural genes coding for enzyme functions, sex-ratio distorter factors and developmental proteins. Most of the mapped biochemical loci codify for enzymes that directly control pathways of basic importance to energy metabolism, others control functions that are correlated with these pathways, while some, such as the esterase loci, are involved in specific, not always defined metabolic functions (Table 3). The use of electrophoretic variants at these loci has greatly improved the knowledge of medfly population structure,[10,44] comparative genetics, and taxonomy.[45] Two sex-ratio distorter factors, Sd1, Sd2 have been found in the laboratory and in field populations of the medfly.[46,47] These factors are inherited as simple mendelian factors and change the sex-ratio to 5 males : 3 females. Sd1 is closely linked to Est1 in linkage group ap..Pgi (Chromosome 4), Sd2 is linked to Got1 locus in linkage group E.[46,47] The close linkage between the two sex distorter genes between genes and an enzymatic marker might be suitable for exploitation for genetic sexing purposes. In addition, the existence of different autosomal sex factors indicates that the genetic basis of sex determination is less simple than current interpretations imply.

Four independent genes (LspI, LspII, LspIII, LspIV) code for the four major larval serum proteins that show a coordinate, genetically controlled developmental pattern.[48] The recovery of genes that are expressed in a developmental specific fashion could provide useful tools for genetic manipulation.[49]

V. CHROMOSOMAL ABERRATIONS

Translocations have been the most studied chromosome aberrations in the medfly. This was due to: (1) the low number of mutations isolated and described to study other aberrations, and (2) the urgent need for a genetic sexing system (GSS).

Translocations have been isolated by the pseudo-linkage technique[50] by irradiating males with X-rays or gamma irradiation and crossing them with virgin, homozygous mutant females. F1 males were singly backcrossed to female mutants and their progeny scored for pseudo-linkage. A variant to this technique is the irradiation of mutant males instead of the wild-type males, or using multiple marker strains (rather than a single mutant strain) in order to isolate multiple translocations.

By the cytological analysis of Y-autosome translocations, and available mutant markers, specific loci could be assigned to particular chromosomes.[18,19,21] A routine proposed method[38] allows location of any new mutation avoiding the interaction between mutations. It combines three translocations { (T:(Y,2,3), T:(Y,3,5), and T:(Y,4) }, a series of crosses and

Table 3
Biochemical Loci Mapped in the Medfly

Main Metabolic Pathway Enzyme (E. C. number)	Enzyme Symbol	No. of Loci	Locus Symbol	Chromo. Location
Conversion of carbohydrate reserves to glucose-6-phosphate:				
Hexokinase-1 (2.7.1.1)	HK-1		Hk1	5[43]
Hexokinase-2	HK-2	2	Hk2	4[43]
Phosphoglucomutase (2.7.5.1)	PGM	1	Pgm	6[43]
Glycolysis:				
Phosphoglucose isomerase (5.3.1.9)	PGI	1	Pgi	4[43]
Aldolase (4.1.2.13)	ALD	1	Ald	2[33]
Pentose phosphate pathway:				
Glucose-6-phosphate dehydrogen (1.1.1.49)	G6PD	1	Zw	5[43]
6-Phosphogluconate dehydrogenase (1.1.1.44)	6PGD	1	Pgd	5[43]
Alpha-Glycerophospahte suttle:				
Alpha glycerophosphate dehydrogenase (1.1.1.8)	GPDH	1	Gpdh	2[33]
Krebs cycle:				
Isocitrate dehydrogenase(1.1.1.8)	IDH	1	Idh	6[43]
Fumarate hydratase (4.2.1.2)	FH	1	Fh	5[43]
Malate dehydrogenase (1.1.1.37)	MDH-1		Mdh1	3[20]
	MDH-2	2	Mdh2	2[20]
Aconitase (4.2.1.3)	ACON-1	1	Acon1	2[20]
Others:				
Xanthine dehydrogenase (1.2.1.37)	XDH	1	Xdh	2[20]
Alcohol dehydrogenase (1.1.1.1)	ADH-1		Adh1	2[20]
	ADH-2	2	Adh2	2[20]
Aldehyde oxidase (1.2.3.1)	AOX	1	Aox	2[20]
Esterases (3.1.1.)	EST-1		Est1	4[43]
	EST-2	4	Est2	4[43]
	EST-6		Est6	6[20]
Beta esterase	B-EST		Est	4[43]
Glut. oxalacetate transaminase	GOT-1		Got1	6[43]
(2.6.1.1)	GOT-2	2	Got2	3[43]
Glut. pyruvate transaminase (2.6.1.2)	GPT	1	Gpt	5[43]
Hydroxiacid dehydrogenase (1.1.1.35)	HAD	1	Had	5[43]
Malic enzyme (1.1.1.40)	ME	1	Me	2[43]
Mannose phosphate isomerase (5.3.1.8)	MPI	1	Mpi	2[20]

Main Metabolic Pathway Enzyme (E. C. Number)	Enzyme Symbol	No. of Loci	Locus Symbol	Chromo. Location
Larval serum proteins:				
Larval serum protein-I	LSP-I		Lsp-I	2[48]
Larval serum protein-II	LSP-II	3	Lsp-II	6[48]
Larval serum protein-III	LSP-III		Lsp-III	5[48]
Various:				
Lethal				
Sex ratio distorter 1				
Sex ratio distorter 2				
pentynol resistant				

backcrosses with the mutant that search for pseudo-linkage between sex and the mutation. Their segregation combinations indicate the chromosome location.

The active role of the Y chromosome as a sex determinant in the medfly was confirmed through a study of a T:(Y,2) translocation studied by Zapater and Robinson.[19] Thus, the medfly differs from *Drosophila* where the Y chromosome has no role in the determination of sex. This translocation involves an insertion of a large fragment of the distal part of the long arm of the Y chromosome into chromosome no. 2[51] and the individuals carrying that chromosome are males regardless of the X or XX possible genotype complements.

Translocations have been a powerful "tool" in the construction of genetic sexing strains (GSS). The first such strain consisted of brown-pupa (wild-type) males and dark-pupa females.[52] An electro-optical sorting machine separated the sexes at the pupal stage. A similar pupal sexing system was constructed by Robinson and van-Heemert,[53] with white-pupa females and brown-pupa (wild-type) males.

Other GSS were constructed with sex-limited phenotypes such as allyl-alcohol susceptibility[54] or the rosy phenotype with purine susceptibility.[27] These were tested on a limited scale in the laboratory but never reached the field testing stage. The allyl-alcohol dependent GSS utilized an alcohol dehydrogenase null mutant induced with X-rays.[55] The null mutant was later cytologically confirmed to be a chromosomal deletion[56] and was the first described in that species.

The only strain types that reached the practical field testing stage was the GSS based on the white-pupa mutant. Various additional white-pupa GSS were developed.[57,58]

The scheme presented in Figure 1 shows how GSS operates. Aneuploid individuals generated from adjacent-I segregation (and eventually adjacent-II segregation) frequently die in early developmental stages, generating a reduced offspring in the strain. The alternate segregation

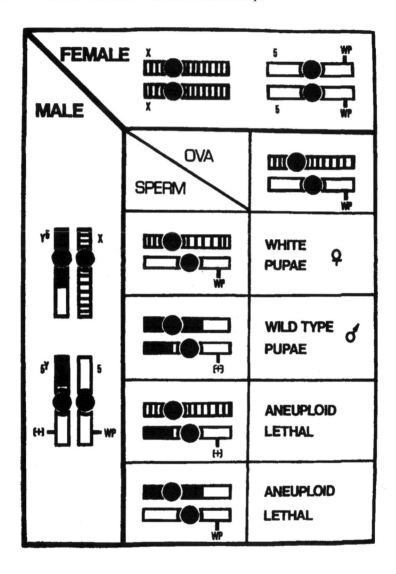

Figure 1. Scheme for the assay of a T:(Y,A) translocation by pseudo-linkage in *Ceratitis Capitata.* Males emerge from brown pupae and females from white pupae.

euploid males and females produce offspring with parental genotypes perpetuating the system, provided that contamination and recombinations do not occur.

Male recombination interferes with the GSS, producing white-pupa males and brown-pupa females, with increasing frequency in subsequent generations due to selection. Hence the presence of recombination in males should be monitored for 10 to 14 generations in a candidate strain before it is mass-reared for GSS. The presence of low-level male recombination in strains constructed to-date necessitated their periodical replacement. It could be possible to stabilize such strains by the isolation of inversion, which suppress chromosomal recombination. The substantial linkage maps which are available to-date[25,26] enable such attempts to be carried out.

The isolation of duplication/deficiencies in this species was first described by Robinson[59] who studied the T:(Y,2) strain. The data were based on survival values of various life stages and pupa-adult segregation. One of the aneuploid genotypes (adjecent-I) survived to the pupal stage. Similarly, adjacent-I segregation males, were reared to the adult stage in a T:(Y,3) strain.[36] These males could mate but were sterile. The model was confirmed by cytological analysis showing that a large portion of the long arm of the Y chromosome was missing in the males.[51] Adjacent-I segregation products that survived until adulthood were also confirmed to appear in Y-linked translocations;[60] this negative impact analyzed, and consequently proposed to have GSS translocation breakpoints not only close to the GS locus, but also close to the middle of the chromosome. This requirement should result in more stable GSS.

VI. CHROMOSOMAL MAPS

A. CHROMOSOME 2

Five morphological and 11 biochemical loci have been described on that autosome. The two biochemical genes Adh1 and Adh2 which are responsible for the ADH system in the medfly[20] are closely linked (0.49 cM) on this autosome. By deletion mapping the Adh region has been cytologically located near the free end of the left arm of the second chromosome within an area between the 2C and 3A segments. Genetic analysis of the region around Adh has identified seven neighboring biochemical genes (Acon1, Mpi, Est6, Aox, Mdh2, LspI). The orientation of loci with regard to the centromere sets the origin of the map of the left arm of the second chromosome close to the two Adh loci. Preliminary studies showed the Adh and the Oe loci to be approximately 36 to 45 map units apart (as calculated from female recombination data).[61]

B. CHROMOSOME 3

Three morphological loci and two biochemical loci have been described on that chromosome. Recombination data are available for the dark-pupa (dp) and eroded-wings (ew) loci only. Recombination in females between ew and dp is 21.57% and in males 0.14%.[25]

Table 4
Chromosome No. 4 - Linear Map of Loci and Recombination Values

Locus	ap	B	sb	dc	db	Hk2	lt	Sd1	Est1	Est2	Pgi	
ap	xxxxx	0.3	5.7	16.3								
B	0.0	xxxxx	2.9	10.5	26.1		47.6					
sb	0.3	0.1	xxxxx	10.5	18.3		48.6					F
dc	0.0	0.0	1.4	xxxxx	6.5	8.8	44.3					E
db	0.0	0.0	2.5	0.7	xxxxx							M
Hk2						xxxxx		28.8	28.8	28.8	40.2	A
lt		0.1	1.5	0.1			xxxxx					L
Sd1								xxxxx	0.0	0.0		E
Est1									xxxxx			S
Est2										xxxxx		
Pgi											xxxxx	

RECOMBINATION IN MALES

Table 5
Chromosome No. 5 - Linear Map of Loci and Recombination Values

Locus	Cy	rb	we	or	Gpt	ye	Hk1	ro	Fh	wp	Zw	
Cy	XXX	13.0		26.0		44.4		43.8		44.8		
rb	9.5	XXX	5.9	30.3		42.2		44.4		49.5		F
we		0.0	XXX			37.9	47.2	45.3		48.4		E
or	9.0	0.0		XXX	20.0	20.6		28.9		47.4		M
Gpt					XXX		8.1					A
ye	8.4	0.1	0.0	0.3		XXX		11.3		39.9		L
Hk1							XXX		19.7			E
ro	5.9	0.4	0.0	0.1		0.6		XXX		34.6		S
Fh									XXX	32.1		
wp	7.0	0.4	0.4	0.7		0.1		0.1		XXX	34.3	
Zw											XXX	

RECOMBINATION IN MALES

C.   CHROMOSOME 4

Eleven morphological loci and six biochemical loci were described on that chromosome. Recombination levels between 11 of them were ascertained (Table 4) and their linear order is as follows: ap, B, sb, dc, db, Hk1, lt, Sd1, Est1, Est2, Pgi.

Low, but persistent recombination was also recorded in males between some of the loci studied. The first case of male recombination in this species was recorded between the ap and dc loci on that chromosome.[40] Additional studies showed male recombination to be common although at relatively low levels in all the chromosomes studied. The highest level of recombination in males (on the fourth chromosome) was found between the sb and db loci (2.5%). We have also recorded recombinations involving the Sr and Sp loci, but the data were very erratic, showing similar and relatively high recombination levels in males and females especially where the Sr locus was involved.[25] It was impossible to ascertain their relative location on the third chromosome.

D.   CHROMOSOME 5
Thirteen morphological and seven biochemical and functional loci were described on that chromosome. Recombination levels between 11 of them were ascertained (Table 5) and their linear order is as follows: Cy, rb, we, or, Gpt, ye, Hk1, ro, Fh, wp, Zw. Male recombination was also recorded on that chromosome, and the highest levels involved the Cy locus, when tested with all the available morphological loci.

E.   CHROMOSOME 6
A direct test to correlate this chromosome and the linkage group E[10] has yet to be performed, but by the process of elimination, the gene cluster E can be assigned to chromosome 6. Five morphological and four biochemical loci were described on that chromosome. Recombination levels were studied only for the gr and bo loci, which amount to 11.1% in females and 3.8% in males. The bo and fs have recombination levels between 4 - 5%.[62]

1.   Sex Chromosomes
Only a single morphological gene was mapped so far on the X chromosome, the rosy ry mutant which is recessive and affects the eye color as well as being purine susceptible.

VII. UTILIZATION OF GENETIC INFORMATION IN SIT PROJECTS

A.   GENETIC SEXING
The study of the genetics of the medfly resulted from practical needs created by the implementation of the Sterile-Insect-Technique (SIT) of that pest. To enhance the utilization of SIT and improve it's efficacy for various pest species,[63] it was necessary to develop methods of "male release" only, hence the need to construct a method of sex separation of the flies in the mass production process. Sex-sorting or sex-killing methods have been developed in the past for various insects (beneficial and injurious).[64,65]

Since the mid 1970s, the Insect and Pest Control Section of the FAO/IAEA Joint Division concentrated efforts to devise practical methods of obtaining pure medfly male batches for release in SIT projects. It was necessary in regions where fresh fruit is being exported, to consider "sterile stings" caused by the released females that would cull the fruit and render it unmarketable, thus undermining the whole project. It was also important for the enhancement of the efficacy of the released males and the elimination (or at-least reduction) of mating between the released (sterile) males and females.

A better insight into the genetics of that species was a prerequisite to the development of such "genetic sexing" methods. Sex separation based on sex-limited (or sex-linked) traits, required that such traits be isolated or engineered and that the chromosomes of the insect be marked and be readily recognized.

The first "genetic sexing" strain was constructed in 1977, and utilized the dark-pupa locus dp which is autosomal and recessive. It was translocated to the male's Y-chromosome by irradiation. Two such strains were developed (line 69 and line 127) which consisted of brown-pupa (wild-type) males and dark-pupa females.[52] These lines were than followed by other lines which utilized the white-pupa wp recessive autosomal locus.[53,56] The later lines were used in actual SIT cage and field tests with reasonable success in Italy[66] and Israel.[67] In these tests, the separation of males from females was carried out by commercial electro-optical sorting machines, adapted for sorting medfly pupae.

"Sex-killing" (or the differential killing of the female gender) utilizing conditional lethals was the ultimate goal of most studies of genetic sexing. Numerous avenues were investigated, such as the use of alcohol dehydrogenase,[54] use of temperature conditional lethals,[68] dieldrin resistance,[69] purine sensitivity of the X-chromosome rosy locus,[27,70] etc. None of these have reached the stage where it could be applied and field tested.

The design of the genetic sexing strain is based on the construction of a translocation between the male's Y-chromosome and an autosomal piece carrying the discriminating alleles. As stated above, a low but persistent level of recombination exists in males. This will result in a slow but steady decay of the genetic sexing strain, due to leakage of the discriminating allele between males and females.

Stabilization of such strains could be and have been obtained by the construction of crossing-over suppressors.[27,71,72]

VIII.   CONCLUSION

Genetic research of the Mediterranean fruit fly has been reported from 14 laboratories in 1987 (unpublished data). Some of them showed some degree of coordination through the hospices of the FAO/IAEA coordinated research program on SIT. Others had little or no contact with each other. The very low flow of information between laboratories and the rapidly accumulated information in the various laboratories is a sure prescription for chaos in the very near future. That situation has been discussed extensively in the last decade, but very little has been done to remedy the problem.

There is an urgent need to consolidate the information and weed out redundancies. The informal "Genetic Information Circular" issued by the FAO/IAEA was a significant step but it appeared only up to 1990.

It is necessary to establish a mutant stock center or at least devise active exchange of genetic material between laboratories. This will enable the rapid assembly of basic information on the mutants and their relation to general available information.

It is necessary to update the list of laboratories that are actively involved in genetic studies of the medfly and list their activities and availability of medfly stocks.

A major obstacle to such coordinated efforts is the fact that these studies involve an extremely injurious pest and the exchange of live material is prevented by strict quarantine regulations. There is no simple way to overcome this, and it seems that if stock centers are established they should be located in countries that have no such restrictions, or where the medfly could not become a pest due to climatic conditions.

Such activities require a modest amount of funding and, in the present state of national funding of research, it seems that only a multinational or international organization could carry that burden. If SIT would be the future method of medfly control, and recent field tests showed that it is not only possible but also very realistic, then genetic studies of the medfly should be recommended and assisted in all possible ways.

REFERENCES

1. Sharp, J. L. and Chambers, D. L., A white-eye mutant of the Mediterranean fruit fly, *J. Econ. Entomol.*, 66, 560, 1973.
2. Cavicchi, S., An orbital bristle mutant in *Ceratitis capitata* Wied., *Genetica Agraria*, 27, 204, 1973.
3. Cavicchi, S. and Bellettini, S., Genetic polymorphism of puparium color in *Ceratitis capitata* Wied., *Monitore Zool. Ital.* (N. S.), 7, 135, 1973.
4. Anonymous, Medfly Genetics Information Circular, FAO/IAEA, Circular, Vienna, Austria, 9 pp., 1985.
5. Anonymous, Second Genetics Information Circular, FAO/IAEA, Circular, Vienna Austria, 11 pp, 1987.
6. Anonymous, Genetic Sexing of the Mediterranean Fruit Fly, FAO/IAEA, STI/PUB/828, Vienna. Austria, 224 pp, 1990.
7. Rossler, Y., Genetic sexing of insects by phenotypic characteristics, with special reference to the Mediterranean fruit fly, in *Sterile insect technique and radiation in insect control*, IAEA/STI/PUB/595, Vienna, Austria, 1982, 291.
8. Rossler, Y., Genetic maps and markers, in *World Crop Pests, Fruit Flies, Their Biology, Natural Enemies and Control*, Vol. 3B, Robinson A. S. and Hooper, G. H. S., Eds., Elsevier Sci. Publ., Amsterdam, 1989, 13.
9. Saul, S. H, Genetics of the Mediterranean fruit fly (*Ceratitis capitata*) (Wiedemann), *Agric. Zool. Rev.*, 1, 73, 1986.

10.   **Saul, S. H. and Rossler, Y,** Genetic markers of the autosomal linkage groups of the Mediterranean fruit fly, *Ceratitis capitata*, (Diptera: Tephritidae), *Ann. Entomol. Soc. Amer.*, 77, 323, 1984.

11.   **Bush, G. L.,** The cytotaxonomy of the larvae of some Mexican fruit flies in the Genus *Anastrepha* (Diptera: Tephritidae), *Psyche*, 69, 83, 1962.

12.   **Radu, M., Rossler, Y., and Koltin, Y.,** The chromosomes of the Mediterranean fruit fly, *Ceratitis capitata* (Wied.): Karyotype and chromosomal organization, *Cytologia*, 40, 823, 1975.

13.   **Bedo, D. G.,** Polytene and mitotic chromosome analysis in *Ceratitis capitata* (Diptera: Tephritidae), *Canad. J. Genet. and Cytol.*, 28, 180, 1986.

14.   **Bedo, D. G.,** Polytene chromosome mapping in *Ceratitis capitata* (Diptera: Tephritidae), *Genome*, 29, 598, 1987.

15.   **Gasperi, G., Malacrida, A. L., Milani, R., Rubini, P. G., and Sacchi, L.,** Genetical studies on *Ceratitis capitata* Wied. - progress report, in *Fruit Flies of Economic Importance*, Cavalloro, R., Ed., A. A. Balkema Publ., Rotterdam, 1983, 148.

16.   **Southern, D. I.,** Cytogenetic observations in *Ceratitis capitata*, *Experientia*, 32, 20, 1976.

17.   **Zacharopoulou, A.,** Cytogenetic analysis of the mitotic and salivary gland chromosomes in the medfly, *Ceratitis capitata*, *Genome*, 29, 67, 1987.

18.   **Zacharopoulou, A.,** Polytene chromosome maps in the medfly, *Ceratitis capitata*, *Genome*, 33, 184, 1990.

19.   **Zapater, M. and Robinson, A. S.,** Sex chromosome aneuploidy in a male-linked translocation in *Ceratitis capitata*, *Canad. J. Genet. and Cytol.*, 28, 161, 1986.

20.   **Malacrida, A. R., Gasperi, G., Zacharopoulou, A., Torti, C., Riva-Francos, E. R., and Milani, R.,** Evidence for a genetic duplication involving alcohol dehydrogenase genes in *Ceratitis capitata*, *Biochem. Genet.*, 30, 35, 1992.

21.   **Busch-Petersen, E. and Southern, D. I.,** Induced suppression of genetic recombination in females of the Mediterranean fruit fly, *Ceratitis capitata* (Wied.), by translocation heterozygosity, *Genetica*, 72, 161, 1987.

22.   **Rossler, Y.,** The genetics of the Mediterranean fruit fly: A "white-pupa" mutant, *Ann. Entomol. Soc. Amer.*, 72, 583, 1979.

23.   **Rossler, Y. and Koltin, Y.,** Fitness of the "apricot-eye" mutant of the Mediterranean fruit fly, *Ceratitis capitata*, *Ann. Entomol. Soc. Amer.*, 70, 544, 1977.

24.   **Rossler, Y. and Rosenthal, H.,** Genetics of the Mediterranean fruit fly (Diptera: Tephritidae): Eye color, eye shape and wing mutations, *Ann. Entomol. Soc. Amer.*, 81, 350, 1988.

25.   **Rossler, Y. and Rosenthal, H.,** Genetics of the Mediterranean fruit fly, *Ceratitis capitata* Wied. as a tool in the Sterile-Insect-Technique, in *Genetic sexing of the Mediterranean fruit fly*, FAO/IAEA, STI/PUB/828, Vienna, Austria, 1990, 69.

26.   **Rossler, Y. and Rosenthal, H.,** The genetics of the Mediterranean fruit fly, *Ceratitis capitata* (Diptera: Tephritidae): Morphological mutants on chromosome five, *Ann. Entomol. Soc. Amer.*, 1992 (In Press).

27.   **Saul, S. H.,** Rosy-like mutant of the Mediterranean fruit fly, *Ceratitis capitata* (Diptera: Tephritidae), and its potential for use in a genetic sexing program, *Ann. Entomol. Soc. Amer.*, 75, 480, 1982.

28.   **Saul, S. H.,** Diagnosing the heterogamic sex in the Mediterranean fruit fly, (Diptera: Tephritidae): The first sex linked gene, *Ann. Entomol. Soc. Amer.*, 78, 198, 1985.

29.   **Saul, S. H.,** Two new eye color mutants in the Mediterranean fruit fly, *Ceratitis capitata*, *Proc. Hawaii Entomol Soc.*, 25, 125, 1985a.

30.   **Carante, J. P.,** Genetique de la mouche Mediterranee de fruits, *Ceratitis capitata* Wied.: Un mutant a oeil fluorescent, *C. R. Acad. Sci. Paris*, 292, 17, 1981.

31. **Carante, J. P.,** Genetics of the Mediterranean fruit fly, *Ceratitis capitata* (Diptera: Tephritidae), a 'reflectionless' eye mutant, *Ann. Entomol. Soc. Amer.,* 75, 613, 1982.

32. **Manso, F. and Lifschitz, E.,** Dos mutantes morphologicas en la mosca del Mediterraneo, *Ceratitis capitata* (Wied.), *Bol. Genet. Inst. Fitotech,* 10, 35, 1979.

33. **Malacrida, A. R.,** unpublished data.

34. **Saul, S. H.,** unpublished data.

35. **Rossler, Y.,** unpublished data.

36. **Zapater, M. and Battista, M.,** Yellow: a new mutant in *Ceratitis capitata, Fruit Flies: Biology and Management,* Aluja, M. and Liedo, P., Eds., Springer-Verlag, New York, 1993, 89.

37. **Rossler, Y. and Koltin, Y.,** The genetics of the Mediterranean fruit fly, *Ceratitis capitata*: Three morphological mutations, *Ann. Entomol. Soc. Amer., 69, 604, 1976.*

38. **Zapater, M.,** Easy method for chromosome location of new mutations in the medfly, *Ceratitis capitata,* Manuscript.

39. **Martinez-Rey, C. E. and Zapater, M.,** Neuvos mutantes en la mosca del Mediterraneo, *Ceratitis capitata,* II Congresso Argentina de Entomologhia, La Cumbre, Dec. 3-6, 1991, 32.

40. **Rossler, Y.,** Recombination in males and females of the Mediterranean fruit fly (Diptera: Tephritidae) with and without chromosomal aberrations, *Ann. Entomol. Soc. Amer.,* 75, 619, 1982a.

41. **Lifschitz, E.,** unpublished data.

42. **Robinson, A. S.,** unpublished data.

43. **Malacrida, A. R., Gasperi, G., Baruffi, L., and Milani, R.,** The contribution of formal genetic studies to the characterization of the Mediterranean fruit fly, *Ceratitis capitata,* in *Genetic Sexing of the Mediterranean Fruit Fly,* FAO/IAEA STI/PUB/828, 1990, 85.

44. **Gasperi, G., Guglielmino, C. R., Malacrida, A. R., and Milani, R.,** Genetic viability and gene flow in geographic populations of the medfly, *Ceratitis capitata* (Wied.), *Heredity,* 67, 347, 1991.

45. **Malacrida A. R., Guglielmino, C. R., Gasperi, G., Baruffi, L., Vilani, P. C., and Milani, R.,** Genetical approach to systematics and phylogeny of Trypetidae (Diptera, Tephritidae), *Boll. Zool.,* 58, 355, 1991.

46. **Milani, R., Gasperi, G., and Malacrida, A.,** Biochemical genetics (of *Ceratitis capitata*), in *World Crop Pests, Fruit Flies, Their Biology, Natural Enemies and Control* (Vol. 3B), Robinson, A. S. and Hooper, G. H. S., Eds., Elsevier Sci. Publ., Amsterdam, 1989, 35.

47. **Malacrida, A. R., Gasperi, G., and Milani, R.,** Genome organisation of *Ceratitis capitata*: Linkage groups and evidence for sex-ratio distorters, in *Fruit flies,* Economopoulous, A. P., Ed., Elsevier Science Publ., Amsterdam, 1987, 169.

48. **Malacrida, A. R., Gasperi, G., Baruffi, L., Biscaldi, G. F., and Milani, R.,** The gene family of larval serum proteins (LSP) in *Ceratitis capitata,* in *Fruit Flies of Economic Importance,* R. Cavalloro, Ed., A. A. Balkema Publ., Rotterdam, 1989, 283.

49. **Anonymous,** Report of Consultant's meeting on the application of genetic engineering and recombinant DNA technology in the development of genetic sexing mechanisms for the Mediterranean fruit fly, *Ceratitis capitata,* FAO/IAEA, Vienna, Austria, 1985Aa.

50. **Muller H. J. and Altenburg, E.,** The frequency of translocations produced by X-rays in *Drosophila, Genetics,* 15, 283, 1930.

51. **Zacharopoulou, A.,** personal communication, 1992.

52. **Rossler, Y.,** Automated sexing of *Ceratitis capitata* (Dip: Tephritidae): The development of strains with inherited, sex-limited pupal color dimorphism, *Entomophaga,* 24, 411, 1979a.

53.   **Robinson, A. S. and van-Heemert, C.,** *Ceratitis capitata*: A suitable case for genetic sexing, Genetica, 58, 229, 1982

54.   **Robinson, A. S., Riva, M. E., and Zapater, M.,** Genetic sexing in the Mediterranean fruit fly, *Ceratitis capitata*, using alcohol dehydrogenase locus, *Theor. Appl. Genet.,* 72, 445, 1986.

55.   **Riva, M. E. and Robinson, A. S.,** Induction of alcohol dehydrogenase null mutants in the Mediterranean fruit fly, *Ceratitis capitata, Biochem. Genet.,* 24, 765, 1986.

56.   **Zacharopoulou, A., Riva, M. E., Malacrida, A., and Gasperi, G.,** Cytogenetic characterization of a genetic sexing strain on *Ceratitis capitata, Genome, 34,* 606, 1991.

57.   **Busch-Petersen, E., Rippel, J., Pyrek, A., and Kafu, A.,** Isolation and mass-rearing of a pupal genetic sexing strain of the Mediterranean fruit fly, in *Modern Methods of Insect Control: Nuclear Techniques and Biotechnology,* FAO/IAEA, STI/PUB/763, 1988, 211.

58.   **Zapater, M.,** Two new pupal sexing strains in the Mediterranean fruit fly, *Ceratitis capitata,* in *Genetic Sexing of the Mediterranean Fruit Fly,* FAO/IAEA, STI/PUB/828, Vienna, Austria, 1990, 107.

59.   **Robinson, A. S.,** Unexpected segregation ratios from male-linked translocations in the Mediterranean fruit fly, *Ceratitis capitata, Genetica,* 62, 209, 1984.

60.   **Kerremans, P. H., Gencheva, E., and Franz, G.,** Genetic and cytogenetic analysis of the Y-autosome translocation in the Mediterranean fruit fly, *Ceratitis capitata, Genome,* 35, 264, 1992.

61.   **Riva, M. E.,** personal communication, 1992.

62.   **Martinez-Rey, C. E.,** personal communication, 1992.

63.   **Calkins, C. O.,** The effect of mass-rearing on mating behavior of the Mediterranean fruit flies, in *Proc. Int'l Symp. Biol. and Control Fruit Flies,* Kawasaki, K., Iwahashi, O., and Kaneshiro, K., Ginowan, Japan, 1991, 153.

64.   **Davidson G.,** *Genetic Control of Insect Pests,* Academic Press, London and New York, 1974, 158 pp.

65.   **Pal, R. and Whitten, M.,** *The Use of Genetics in Insect Control,* Elsevier, North Holland, 1974, 241 pp.

66.   **Robinson, A. S., Cirio, U., Hooper, G. H. S. and Caparella, M.** Field cage studies with a genetic sexing strain in the Mediterranean fruit fly, *Ceratitis capitata, Entomol. Exp. and Appl.,* 41, 231, 1986.

67.   **Nitzan, Y. and Rossler, Y.,** unpublished data.

68.   **Busch-Peterson, E.,** personal communication.

69.   **Busch-Petersen, E. and Wood, R. J.,** The isolation and inheritance of dieldrin resistance in the Mediterranean fruit fly, *Ceratitis capitata* (Wiedemann)(Diptera: Tephritidae), *Bull. Entomol. Res.,* 1986.

70.   **Saul, S. H.,** Genetic sexing of the Mediterranean fruit fly, *Ceratitis capitata* (Wied.) (Diptera: Tephritidae): Conditional lethal translocations that preferentially eliminate females, *Ann. Entomol. Soc. Amer.,* 77, 280, 1984.

71.   **Busch-Petersen, E. and Kafu, A.,** Assessment of quality control parameters during mass-rearing of a pupal color genetic sexing strain of the Mediterranean fruit fly, *Entomol. Exp. and Appl.,* 51, 241, 1989.

72.   **Busch-Petersen, E. and Kafu, A.,** Stability of two mass-reared genetic sexing strains of *Ceratitis capitata* (Diptera: Tephritidae) based on pupal color dimorphism, *Environ. Entomol.,* 18, 315, 1989.

Chapter 9

REQUIREMENTS AND STRATEGIES FOR THE DEVELOPMENT OF GENETIC SEX SEPARATION SYSTEMS WITH SPECIAL REFERENCE TO THE MEDITERRANEAN FRUIT FLY *CERATITIS CAPITATA*

Gerald Franz and Philippe Kerremans

I. INTRODUCTION

There are several reasons why it is advantageous to release only males during SIT programs. First, it appears that males are the primary active agent in this approach. Therefore, mass rearing only males reduces the program costs. Second, the effectiveness can be increased (or program costs decreased) because the use of only one sex eliminates the problem of preferential mating among the released flies. Third, the monitoring of the release program, which constitutes up to 45% of the overall costs, can be simplified if only males are released. Then the number of females trapped can be used as direct estimation for the wild population size, provided it can be assured that absolutely no females are released. Fourth, in many cases SIT cannot be applied because of the damage caused by the released females (e.g. ovipunctures or sterile stings in the case of fruit flies) or because the females are disease vectors (e.g. mosquitoes and tsetse flies). Whenever the released sterile insects are not harmful, the SIT can be applied as a treatment for control purposes, which is especially advantageous where eradication is impractical or impossible. Fifth, the danger of the accidental release of unsterilized or partially sterilized females is avoided. Releasing fertile males is only detrimental in so far as these flies do not advance the progress of the eradication or control efforts, whereas the accidental release of fertile females could seriously damage the program. To construct a sex separation system successfully for insects, a certain amount of basic knowledge on the genetics and cytogenetics of the respective species is required in addition to information regarding its general biology. The Mediterranean fruit fly *Ceratitis capitata* (Wied.), (medfly) can be used to illustrate: (1) the experimental approaches used to obtain this knowledge, (2) how genetic sexing strains are constructed, and (3) how genetics and cytogenetics can be applied to identify or solve certain problems associated with the construction or maintenance of these strains.

A genetic sexing system as used in the medfly, consists of two principle components: a mutation that can be used as a marker for selection and a Y-autosome translocation linking the wild-type allele of this marker to the male sex. In the following paragraphs, these two components and their experimental manipulation will be described with emphasis on some of those areas that are crucial if a similar system will be developed for other pest insects.

II. PRINCIPAL CONSIDERATIONS

A. LABORATORY REARING OF INSECTS

Because the maintenance of insects under artificial laboratory conditions is a very basic requirement, it has to be established before any genetic experiments can be done. It is essential to establish adequate standard rearing conditions for experimentation that will allow quantitative measurements to be taken without influences from the rearing environment (food, temperature, humidity etc.). Furthermore, it must be possible to cross individual insects, a procedure required for many experimental strategies in genetics, and to maintain several strains in parallel.

B. MUTATIONS

Experimental approaches in genetics depend heavily on mutations, which serve as markers for the respective chromosome and allow them to be followed through genetic crosses. Morphological mutations, visible as alterations on the exterior of the insect, are of special practical relevance. In addition, the penetrance of these mutations should be good, i.e. the variability of the mutant phenotype should be low enough to allow quantitative experiments. Therefore, in each species, mutations must be identified either from wild populations or through induction by chemical mutagens.

A second class of genetic marker is represented by the biochemical mutants. These are in most cases not visible as morphological differences, but can be assayed by biochemical methods such as chromatography. These mutations are especially relevant for investigations aiming at the analysis of whole populations (e.g. isoenzyme variability) because the natural variability of these markers is relatively high.

A third class of mutations is defined by their application in sex separation strains. Depending on the type of selection system (sorting versus killing), they can either be morphological mutants, as described above, or conditional lethal mutations. In the medfly, several mutants that affect the color of the puparium have been used to separate males and females (Table 1). In the class of the conditional lethals, three different mutations have been investigated for their ability to eliminate the females by either addition of the appropriate chemicals to the food or by temperature treatment.

The position of all available mutations relative to each other can be determined on the basis of the recombination frequency. Performing the relevant crosses will lead to the establishment of genetic maps for the linkage groups existing in the respective species. In the medfly two maps exist, one for morphological markers[1] and one for enzyme markers.[2]

TABLE 1
MUTATIONS USED AS SEXING GENES IN MEDFLY

Chromosome	Name	Phenotype
2	niger (<u>nig</u>)	dark puparium[3]
2	rosy eye (<u>ry</u>)	purine sensitive[4]
2	Alcohol dehydrogenase (<u>Adh</u>)	alcohol sensitive[5]
3	dark pupae (<u>dp</u>)	dark puparium[6]
5	white pupae (<u>wp</u>)	white puparium[7]
5	temperature sensitive sensitive lethal (<u>tsl</u>)	temperature[8] sensitive

C.   CHROMOSOME ABERRATIONS

Chromosome rearrangements are usually induced by irradiation, e.g. gamma radiation.   In the context discussed here, they can be used as experimental tools to determine the chromosome on which a particular mutation is located (e.g. translocations) or to map individual mutations on the respective chromosome (e.g. deletions or transpositions).   This can be achieved by combining genetic data with the analysis of polytene chromosomes (see below) and allows the orientation and the alignment of the genetic linkage map with the chromosomes to be determined.

In addition, translocations are one of the essential components of all existing sexing strains in the medfly.   All mutations that are used either to separate the sexes or to eliminate one of them, are not in themselves sex-specific.   Genetic sexing can only be achieved by linking the respective wild-type allele via translocations to the chromosome that is responsible for the male sex determination (the Y-chromosome in medfly).[9]

D.   CYTOLOGY

In the first step, the optimal tissues for the analysis of mitotic, meiotic or polytene chromosomes have to be identified following the examples of other species (e.g. *Drosophila*).   Also in many cases, the staining and spreading techniques have to be adapted to the respective system.   The first aim of this approach will be to establish the basics, e.g. chromosome number and shape, numbering system for the chromosomes and a description of the polytene chromosome banding pattern.   This information can then be used to analyze mutant strains, e.g. deletions to map mutations, mapping of translocation breakpoints, etc.

Furthermore, cytogenetic analyses should allow for the establishment of the type of sex determination system that exists in the respective species. This is an absolutely essential prerequisite for the construction of genetic sexing strains because the selectable marker gene has to be linked to sex.

III. APPLICATION OF GENETICS AND CYTOGENETICS FOR THE CONSTRUCTION OF SEX SEPARATION STRAINS IN THE MEDFLY

A. A TEMPERATURE SENSITIVE LETHAL MUTATION AS SEXING GENE

The separation of the sexes on the basis of pupal color has several disadvantages. The most important drawback concerns the equipment required for sorting. It is very expensive, not completely accurate and can damage the pupae. The conditional lethal mutations, rosy eye and Adh, on the other hand require the application of either very expensive (purine) or toxic (allyl alcohol) chemicals. Therefore, alternatives were investigated. The utilization of temperature sensitive lethal mutations seemed very appropriate because it is known from *Drosophila* that many of these mutations exist, and provided the correct screening procedure is used, it should be possible to identify mutations that allow elimination of the females very early during development and with relatively cheap, but accurate equipment.

One temperature sensitive lethal mutation has been identified in the medfly.[8] In a genetic sexing strain, this mutation allows elimination of the females as embryos or early stage larvae if these stages are treated with elevated temperatures. A second phenomenon associated with this mutation is a delay in development at sub-lethal temperatures. This alone is sufficient to separate the wild-type males from the mutant, i.e. delayed females.

B. GENETIC STABILITY OF SEX SEPARATION STRAINS

The second component of a sexing strain, besides the selectable marker, is the translocation that links the autosome carrying the wild-type allele of this marker to the male determination chromosome, i.e. in medfly, the Y-chromosome (Figure 1). Because the males are heterozygous for the sexing gene, genetic recombination between this gene and the translocation breakpoint will destroy the sex linkage (Figure 2). Although the recombination frequency in males is very low, selection during mass-rearing can increase the relative number of the recombinants if these are more viable than the non-recombinant flies. This can, eventually, lead to the complete breakdown of the sexing system (Figure 3).

Utilizing the existing mutations[1] and the cytology developed for the medfly,[10,11] improved sexing strains were constructed to minimize this problem. The underlying strategy was to bring the translocation breakpoint and the selectable marker close together on the chromosome, because the distance between them determines the recombination frequency. To achieve this, two separate lines of experiments were required. First, the location of the respective sexing gene on the polytene chromosome had to be determined.

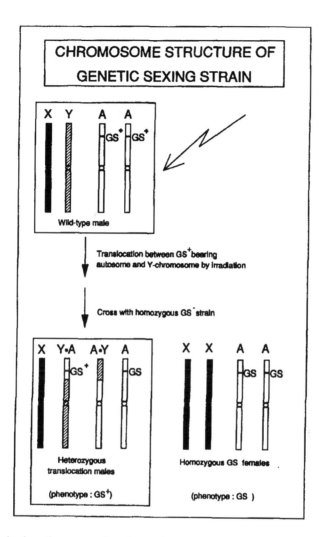

Figure 1. A schematic representation of a genetic sexing strain. Two principal components are required: a selectable marker (GS⁺ = dominant wild-type allele, GS = recessive mutant allele) and a Y-autosome translocation linking the wild-type allele to the male sex. In the resulting strain, the males are heterozygous and the females are homozygous for the sexing gene.

Mapping of the temperature sensitive lethal, the white pupae and other mutations has been achieved by translocation, deletion (Figure 4), and transposition mapping.[12,13] A second group of experiments was aimed at the induction, isolation and analysis of several new translocations between chromosome 5 that carries the temperature sensitive lethal and the

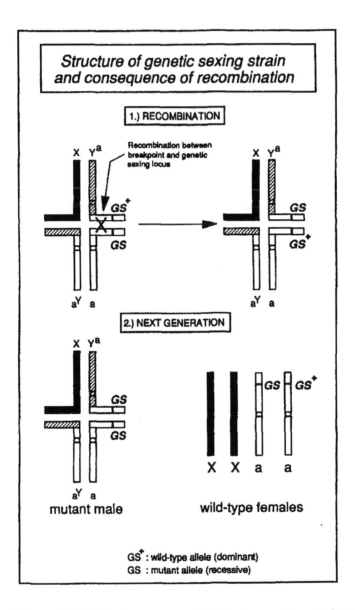

Figure 2. Recombination in the heterozygous males in the chromosome segment between the translocation breakpoint and the selectable marker will lead to breakdown of the sexing system, i.e. mutant males and wild-type females will occur.

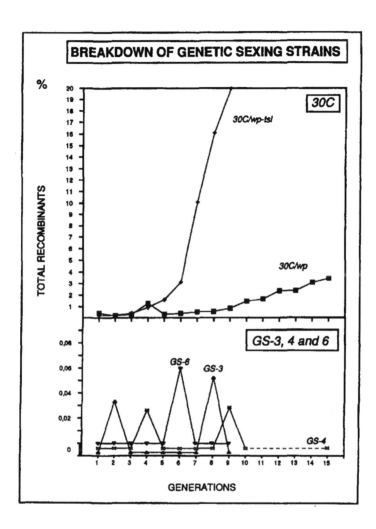

Figure 3. If recombinants cannot be removed from a mass reared colony, they will, depending on their fitness, accumulate in the colony. In the upper part, the Y-autosome translocation T(Y;5)30C is shown either with wp or with wp-tsl as sexing genes. When these strains are reared without removing aberrant flies after each generation, a steady increase of recombinant phenotypes is observed. In the lower part, the same test is shown for three new Y-autosome translocation strains. They differ from T(Y;5)30C in the location of the translocation breakpoint on chromosome 5, i.e. their breakpoints are much closer to wp and tsl.

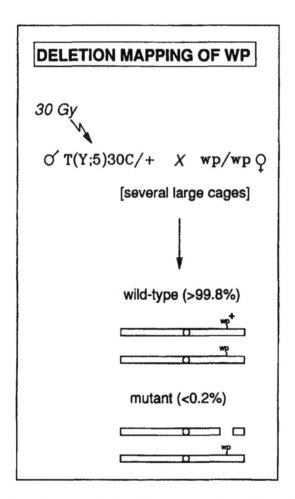

Figure 4. A schematic representation of the deletion mapping of <u>wp</u> is shown. As there are no balancer chromosomes available in medfly, the deletions had to be induced in a Y-autosome translocation.

Y-chromosome (Figure 5). Through the analysis of polytene and mitotic chromosomes and through genetic crosses, the structure of these new translocations was determined. This allows one to select the most suitable strains before they are transferred to the large scale mass rearing program.

Knowing the chromosomal location of the sexing genes allowed one to select those translocations that have the breakpoint close to them. The resulting new strains are clearly much more stable, i.e. they produce recombinants at a frequency below 0.1% and do not show an accumulation of recombinants (Figure 3).[14]

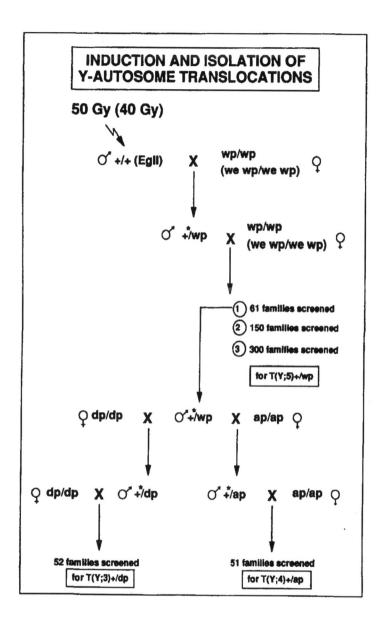

Figure 5. The strategy used to isolate new Y-autosome translocations is shown. In three separate experiments, 511 males were screened for pseudolinkage between markers located on chromosome 5 and the Y-chromosome. In addition, approximately 50 males were screened with apricot eye (<u>ap</u>) and dark pupae (<u>dp</u>) located on chromosome 4 and 3, respectively.

REFERENCES

1. **Rössler, Y,** Genetic maps and markers, in *World Crop Pests: Fruit Flies, Their Biology, Natural Enemies and Control,* Vol. 3B, Robinson, A. S. and Hooper, G., Elsevier, Amsterdam, 1989, 13.

2. **Milani, R., Gasperi, G., and Malacrida, A.,** Biochemical genetics, in *World Crop Pests: Fruit Flies, Their Biology, Natural Enemies and Control,* Vol. 3B, Robinson, A. S. and Hooper, G., Elsevier, Amsterdam, 1989, 33.

3. **Manso, F. and Lifschitz, E.,** Two morphological mutations found in the Mediterranean fruit fly *Ceratitis capitata, Bol. Genét. Inst. Fitotec. Castelar,* 10, 31, 1979.

4. **Saul, S. H.,** Rosy-like mutant of the Mediterranean fruit fly, *Ceratitis capitata* (Diptera: Tephritidae), and its potential for use in a genetic sexing program, *Ann. Entomol. Soc. Am.,* 75, 480, 1982.

5. **Riva, M. E. and Robinson, A. S.,** Induction of alcohol dehydrogenase (<u>Adh</u>) null mutants in the medfly, *Ceratitis capitata, Biochem. Genet.,* 17, 765, 1986.

6. **Rössler, Y. and Koltin, Y.,** The genetics of the Mediterranean fruit fly *Ceratitis capitata*:: three morphological mutations, *Ann. Entomol. Soc. Am.,* 69, 604, 1976.

7. **Rössler, Y.,** The genetics of the Mediterranean fruit fly: a "white pupae" mutant, *Ann. Entomol. Soc. Am.,* 72, 583, 1979.

8. **Franz, G. and Busch-Petersen, E.,** Analysis of a temperature sensitive lethal mutation used as selectable marker in a genetic sexing strain of the Mediterranean fruit fly *Ceratitis capitata* (In preparation).

9. Lifschitz, E. and Cladera, J. L., Cytogenetics and sex, in *World Crop Pests: Fruit Flies, Their Biology, Natural Enemies and Control,* Vol. 3B, Robinson, A. S. and Hooper, G., Elsevier, Amsterdam, 1989, 1.

10. **Bedo, D. G.,** Polytene chromosome mapping in *Ceratitis capitata* (Diptera: Tephritidae), *Genome,* 29, 589, 1987.

11. **Zacharopoulou, A.,** Polytene chromosome maps in the medfly *Ceratitis capitata, Genome,* 33, 184, 1990.

12. **Kerremans, Ph. and Franz, G.,** Cytogenetic map of chromosome 5 of the medfly *Ceratitis capitata* (In preparation).

13. **Kerremans, Ph., Gencheva, E., and Franz, G.,** Genetic and cytogenetic analysis of Y-autosome translocations in the Mediterranean fruit fly, *Ceratitis capitata, Genome,* 35, 264, 1992.

14. **Franz, G., Gencheva, E., and Kerremans, Ph.,** Improved stability of genetic sex-separation strains for the medfly, *Ceratitis capitata* (In preparation).

Chapter 10

FRUIT FLY PROBLEMS IN CHINA AND PROSPECTS FOR USING THE STERILE INSECT TECHNIQUE

Hua-song Wang and He-qin Zhang

I. INTRODUCTION

China has a diversity of geographical environments including temperate, subtropical, and tropical zones that allow for production of a great variety of fruits and vegetables that are hosts for tephritid fruit flies. About 400 species of fruit flies have been recorded from China.[1] Except for *Dacus tau, Bactrocera cucurbitae, Bactrocera dorsalis,* and *Dacus citri,* the hosts of the other species are still unknown.

In the last decade, China has made great strides in economic progress. The people's living standard and purchasing power increased and their demand for vegetables and fruits became much greater. To meet the increasing demand for fruits, large areas have been devoted to fruit production, especially for growing citrus. The area of citrus orchards increased to 1 million hectares in 1989 and the annual output of citrus fruit increased from 0.9 million tons in 1980 to 4.56 million tons in 1989. A great deal of effort has been exerted for fruit protection following the expansion of the areas of citrus production and damage by fruit flies has been reduced. However, the loss caused by fruit flies is still heavy.

II. FRUIT FLY PESTS

A. *BACTROCERA CUCURBITAE*

Bactrocera cucurbitae is found in south and southwest China. Its preferred hosts are bitter melon, cucumber, and *Luffa cylindrica.* There are 4-5 generations a year in the Guangzhou area. The larvae can be found commonly in one of its hosts, *Luffa cylindrica,* in December. During the 1960's, *B. cucurbitae* was an important pest on some vegetables. However, in recent years the infestation by this pest was reduced, probably due to the increased use of insecticides in commercial vegetable farms. Bitter melon is still seriously attacked by this pest in backyard gardens where insecticides are seldom used.

B. *DACUS TAU*

Dacus tau is widely distributed in south and southwest China. Its hosts include cucurbitaceous crops, papaya, guava, and carambola. It is the dominant species in melon fields. The biology and ecology of *D. tau* need to be studied in order to find effective ways to control it.

0-8493-4854-4/94/$0.00+$.50

123

Besides the above two species of pests that infest vegetables, two other species--*Zeugodacus caudatus* Fabr. in Sichuan Province and *Chaedacus caudatus nubilis* Hen. in Taiwan Province--also attack cucurbitaceous crops, but information on these species is lacking.

C.   *DACUS SCUTELLATUS*

Dacus scutellatus is called the pomelo fruit fly in China, but so far there is no report of infestations on pomelo. It is widely distributed in the central, east, south, and southwest parts of China. Although it is a common species in south China, the status of its host plants is still unknown.

D.   *BACTROCERA DORSALIS*

Bactrocera dorsalis has been trapped in parts of central and east China and in south and southwest China. There are 3-5 generations a year. Peach, pear, plum, guava, carambola, citrus fruits, and mango are major hosts of this species.

E.   *DACUS TSUNEONIS* MIY

Dacus tsuneonis Miy was only known previously from Taiwan Province but it has since been found in Guanxi Province in 1956 and then in Guizhou Province. It attacks mainly citrus fruits. There is one generation a year. Its biological and ecological characteristics are very similar to *Dacus citri* but the citrus fruits damaged by *Dacus tsuneonis* usually do not rot; the pulps become white, dry, and contracted.

F.   *DACUS CITRI* CHEN

Dacus citri Chen occurs in all of the citrus-producing regions in China except for Zhejiang and Guandong Provinces, ranging climatically from the temperate zone to subtropical and tropical zones. It is highly injurious to sweet orange and other citrus species. In some orchards, the infestation of fruits by this pest has been estimated to amount to more than 50%.[2] During the 1950's, the infestation in some counties in Sichen Province was higher than 80%. Based on the recent report from Huaihua Plant Protection Station in Xunan Province, the infestation in some areas is still as high as 80%. In 1987, the infestation in Bangong Township, Lodian County, Guizhou Province was 22% and in one of these orchards up to almost 100%.

This pest was first recorded in two provinces during the 1940's,[3] and then was found in four other provinces during the 1960's.[4] It is found in even more provinces today. It seems that *D. citri* is able to survive in colder climates and seriously threatens citrus production.

Adults of *D. citri* begin to emerge and mate in late April and their oviposition occurs from the middle of May to the end of July. The infested fruits drop in October, and the larvae leave the fruits for pupation in the soil. The pupal stage takes about 180 days.

III. CONTROL OF *DACUS CITRI* WITH THE STERILE INSECT TECHNIQUE

A program of release of sterile flies for control of the Chinese citrus fly was conducted in Guizhou Province in 1987 and 1989. A mass rearing program was initiated to ensure that enough insects of the proper stage would be available for the SIT program. The artificial larvae diet of *D. citri* consisted of wheat germ powder, carrot powder, sucrose, wheat bran, brewer's yeast, agar, and sorbic acid, etc. Brown sugar and soybean germ powder or hydrolyzed protein are essential to the survival and oviposition of *D. citri.*

A series of tests indicated that there were no significant influences of radiation treatments from 6 krad to 15 krad two days before emergence on emergence rate and life-span of adults (Tables 1 and 2), nor was the mating time changed (Table 3). In fact, when both males and females were exposed to 6 and 9 krad, the mating speed increased significantly. The irradiated sterile adult behavior resembled that of the native population in the field. The sterile mating crosses of 6 krad had some sterility but still suffered from infestations (Table 4). No damage occurred in mating crosses treated with 9 krad. Therefore, a dose of 9 krad of irradiation to pupae two days before emergence was determined as the appropriate sterilizing dose.[5]

The release of sterile *D. citri* was carried out in a Zhonglian orange orchard of about 34 hectares in Huishui County, Guizhou Province of southwest China in 1987 and 1989, respectively. Six to ten evenly distributed sites in the orchard were chosen for release. The pupae were treated with 9 krad of irradiation two days before emergence, then were mixed with sand and fluorescent powder in a container which was covered with a cage. After the female adults were removed, the irradiated sterile male adults, labelled by fluorescent powder, were released in the orchard. About 56,000 and 95,000 irradiated sterile males were released in 1987 and 1989 with the release ratio of sterile to native fruit flies of 12.5:1 and 45:1 in 1987 and 1989, respectively. The rate of damage to oranges by *D. citri* dropped from 7.5% to 0.005%.[6]

Traps were monitored to determine movement and longevity of the sterile flies. The traps were baited with a solution of brown sugar and hydrolyzed protein. The longest dispersal distance of *D. citri* was about 1500 meters in a favorable wind direction during the release program. It seems that the movement and transportation of citrus fruits containing larvae of *D. citri* are the main means of spreading the infestation.

Table 1
Effects of Different Radiation Doses on Mating
Percentages of the Chinese Citrus Fly

Treatments (krad)	Crosses[1]	Mating rates(%)		
		1985	1986	Average
6	S♂xN♀	395	420	408[**][2]
	N♂xS♀	310	300	305[*][2]
	S♂xS♀	330	315	323[*]
9	S♂xN♀	324	305	315[*]
	N♂xS♀	310	315	313[*]
	S♂xS♀	330	375	353[**]
12	S♂xN♀	208	105	157
	N♂xS♀	200	207	204
	S♂xS♀	110	100	105[*]
15	S♂xN♀	132	120	126
	N♂xS♀	180	253	217
	S♂xS♀	100	93	97[*]
Control	S♂xN♀	216	198	207

[1] S = Sterile, N = Normal
[2] * = significantly different from control at P=0.05 (LSD).
   ** = significantly different from control at P=0.01 (LSD).

Table 2
Effect of Different Radiation Doses on Longevity of Chinese Citrus
Fly Adults

Treatments (krad)	Longevity of male (days)			Longevity of female (days)		
	1985	1986	Average	1985	1986	Average
6	35.67	45.96	40.82 a[1]	43.71	50.42	48.87 a[1]
9	38.00	49.38	43.69 a	45.83	50.78	48.32 a
12	37.87	42.51	40.19 a	42.33	43.31	42.82 a
15	38.71	43.29	41.00 a	44.77	45.12	44.95 a
Control	35.42	44.91	40.17 a	32.95	45.73	39.34 a

[1] Means followed by the same letter are not significantly different at P=0.01, LSR.

Table 3
Periodicity of Mating of Irradiated Chinese Citrus Fly

Treatments and Crosses	8:00	9:00	10:00	11:00	12:00	13:00	14:00	15:00	16:00	17:00	18:00	19:00	20:00	Total
6 krad														
S♀ x N♂		2	6	10	18	11	17	15	22	8	7	4		120
N♀ x S♂	2		9	20	17	20	21	22	10	24	15	6	2	168
S♀ x S♂			6	5	18	16	17	14	16	18	9	6	1	126
9 krad														
S♀ x N♂		2	7	8	13	13	11	19	20	13	13	6	1	126
N♀ x S♂		1	6	8	12	14	10	18	17	15	13	7	1	122
S♀ x S♂		2	5	11	21	15	15	19	16	21	15	7	3	152
12 krad														
S♀ x N♂		2	4	8	16	8	8	8	3	3				60
N♀ x S♂			2	3	2	4	6	9	6	5	5			42
S♀ x S♂		2			7	5	3	5	3	3	2			30
15 krad														
S♀ x N♂		1	6	6	10	13	14	18	14	9	8	2		101
N♀ x S♂			3	5	3	11	7	12	3	2	2			48
S♀ x S♂				4	8	5	4	5	2	1	7		1	37
CONTROL		1	2	7	7	13	9	12	7	10	7	3	1	79

Table 4
Relationship Between Irradiation Doses and Sterility Effects

Treatments and Crosses[1]	Total Fruit Numbers	Total Pierced Fruit Numbers	Pierced Fruit Numbers Dissected	From Dissected Fruits		Damaged Fruit Numbers Dissected[2]	Checked Numbers of Damaged Fruits[3]	Damaged Rates %
				Egg Numbers	Larva Numbers			
6 krad								
S♀ x N♂	31	17	12	13	6	7	9.92	32.00
N♀ x S♂	41	20	15	4		5	6.67	16.27
S♀ x S♂	29	23	15	2		1	1.53	5.28
9 krad								
S♀ x N♂	42	24	17			0	0	0
N♀ x S♂	28	9	6			0	0	0
S♀ x S♂	23	15	12			0	0	0
12 krad								
S♀ x N♂	13	8	5			0	0	0
N♀ x S♂	22	13	12			0	0	0
S♀ x S♂	16	3	2			0	0	0
15 krad								
S♀ x N♂	23	8	6			0	0	0
N♀ x S♂	27	15	9			0	0	0
S♀ x S♂	7	0	0			0	0	0

In 1989, about 48,000 irradiated sterile female *D. citri* were released in another small orange orchard of about 1 hectare. The percentage of damaged oranges dropped from 0.55% to 0.003%. Although the release ratio was not obtained, it seemed that releasing sterile females was effective. However, many orange fruits were scarred because the released females still were able to produce oviposition stings.

These preliminary tests indicated that control of *Dacus citri* with the Sterile Insect Technique was feasible. Further release programs will be conducted in all orange orchards of about 100,000 orange trees across Huishui County in the coming five years.

In China, heavy economic losses have been caused by *D. citri, D. tsuneonis, B. dorsalis, B. cucurbitae*, etc. SIT has been used as a biological agent in integrated pest management to eradicate *D. cucurbitae* on many islands in Japan, to eradicate *B. dorsalis* on one island in Japan and to control it in Taiwan. This preliminary SIT test of releasing irradiated sterile flies for control of *D. citri* has indicated that SIT may be an effective control method for fruit flies of economic importance in China.

REFERENCES

1. **Chao, Y. S., and Ming, Y.,** The investigation of fruit flies (Trypetidae-Diptera) injurious to fruits and vegetables in south China, Tech. Bull. Institute of Plant Quarantine Press, Beijing, 1986.
2. **Chen, S. H.,** Two new Dacinae from Szechwan, *Sinensia*, 11(1&2), 131, 1940.
3. **Chen, F. J., and Wong, F. P.,** Study on citrus maggot in Jiangjin County (in Chinese), Agricultural Science, 1, 46, 1943.
4. **Sun, Z .Y.,** The study and control of *Dacus citri* (in Chinese), in *Chinese Plant Protection Science*, Scientific Publishing House, 1961, 912.
5. **Liu, Q. R. et al.,** Effect of radiation on the sterility of Chinese citrus fly, *Acta Agriculturae Nucleatae Sinica*, 6(1), 46, 1991.
6. **Wang, H. S. et al,** Control of Chinese citrus fly *Dacus citri* by male sterile technique, *Acta Agriculturae Nucleatae Sinica*, 5(3), 135, 1990.

Chapter 11

FRUIT FLY PROBLEMS IN SOUTHEAST ASIA AND EFFORTS TO MEET THEM

S. Vijaysegaran

I. INTRODUCTION

The equitable climate of countries in Southeast Asia and the surrounding region enables the cultivation of a diverse range of tropical and temperate fruits. Fruits are also an important component of the diet of the people in this region.[1] Countries like Taiwan, Thailand, Singapore, and Malaysia represents some of the fastest growing economies in the world today. Coupled with this economic growth are changing dietary patterns leading to increased consumption of fruits and vegetables.[2] The region is also home to a variety of exotic fruits that offer the prospect of good export incomes. Countries like Thailand (US $909 million), Philippines (US $348 million), and India (US $106 million) earn considerable incomes from exports of fruits.[1] Malaysia, which has a highly developed plantation industry for rubber and palm oil, is now aggressively pursuing an export orientated fruit industry. Thus the development of horticultural industries to meet domestic consumption as well as those directed for the lucrative export market is an important component of the economic development process of countries in southern Asia.

With the shift from traditional backyard fruit farms to large scale commercial and export oriented production, damages caused by fruit flies are of primary concern. The development and availability of appropriate management strategies and techniques is, therefore, central to the successful expansion of horticultural industries in the region. This is evident in Thailand and the Philippines who have developed postharvest treatments and research programs with the Japanese government for the vapor heat disinfestation of mango for export to Japan. Malaysia is currently working on a similar program.

This chapter will examine the fruit fly problem and the control methods and strategies in current use. Options for the future, particularly the use of the Sterile Insect Technique, will be examined and discussed in light of problems peculiar to the region.

II. SPECIES OF ECONOMIC IMPORTANCE

Southeast Asia and countries in tropical Asia have a rich diversity of plant and animal life. About 160 genera of tephritids are known from this region, including about 180 *Bactrocera* spp. and about 30 *Dacus* spp.[3] This diversity of endemic species presents a different situation when compared to areas where SIT has traditionally been used for control of a single species that usually has been introduced. Fortunately, the number of species causing economic damage are far fewer and can be placed in general groups as follows:

(1)   *Bactrocera dorsalis* complex: includes the oriental fruit fly and seven other closely related pest species, five of which have not been described before (Table 1).

(2)   Melon fly, *Bactrocera cucurbitae* (Coquillett), and other species in the *B. tau* complex that infest a range of cucurbits.

(3)   Peach fruit fly, *B. zonata*: suspected to be a complex of flies.[4] Infests peach, mango, guava, and other fruits.

(4)   Guava fruit fly, *B. correcta*: infests guava, mango, peach, sapodilla, rose apple, and others.

(5)   *B. latifrons* (infests chili), *B. umbrosa* (infests *Artocarpus* spp.), and *B. albistrigata* (infests several *Syzygium* spp).

The oriental fruit fly *Bactrocera dorsalis* (previously *Dacus dorsalis* (Hendel)) and the melon fly have long been recognized as among the world's most damaging tephritids and consequently are of major quarantine importance. The oriental fruit fly was for a long time considered widespread in all tropical Asian countries. However, recent studies on the taxonomy of fruit flies of southern Asia have revealed a vastly different picture of the pest species, their host range and distribution. The *dorsalis* complex comprises flies that are morphologically very similar and often difficult to tell apart under the ordinary light microscope. Drew and Hancock[5] have recorded 52 species in the *dorsalis* complex of fruit flies in tropical Asia. Of these, eight are considered of major economic importance and their geographic distribution is given in Table 1. Perhaps the most significant aspect of this study is the revelation that the oriental fruit fly is not present in Malaysia, Singapore, Brunei, Indonesia or the Philippines as previously thought. Its distribution is more northern and limited to India, Sri Lanka, Myanmar, China, Taiwan, Thailand, Laos, Vietnam, and Kampuchea.

Each country appears to have its own pest species. For example, in Peninsular Malaysia and East Malaysia (Sabah and Sarawak), the carambola fruit fly (designated as Taxon A until formally described) and another species, Taxon B, are the major pest species. They can be separated based on differences in the adults,[6] larvae,[7] and their isoenzymes.[8] Taxon A was reared from 29 hosts (8 cultivated and 21 wild) but breeds primarily in carambola. In unprotected situations, it can cause up to 100% damage to the crop.[9] It is also the species that has spread to Suriname and possibly French

Table 1
Species of Economic Importance in the *Bactrocera dorsalis* Complex of Flies and their Known Distribution[5]

	Cambodia	China	India	Indonesia	Laos	Myanmar	Malaysia West	Malaysia East	Philippines	Singapore	Sri Lanka	Taiwan	Thailand	Vietnam
B. caryae		*									*			
B. dorsalis	*	*	*		*	*				*		*	*	*
B. taxon A			**ᵃ	*			*	*		*		*		
B. taxon B				*			*	*		*			*	
B. spp. (Philippine B)									*					
B. occipitalis								*	*					
B. spp. 1												*		
B. spp. 2														*

ᵃ In the Andaman Islands only.

Guiana in South America[10,11] and is currently known to occur in Malaysia, Thailand, and Indonesia.[12] It tends to predominate in orchard and urban areas and is rarely, if ever, found in undisturbed rain forests.

Taxon B is closely related to the oriental fruit fly. In Peninsular Malaysia it has been recorded on 36 hosts (16 cultivated and 20 wild). It commonly infests papaya, banana, and mango, three important hosts that Taxon A scarcely or never utilizes. Both Taxon A and B have 17 hosts in common (7 cultivated and 10 wild) but in all cases one or the other species predominates quite markedly.[12]

Unlike Taxon A, Taxon B is endemic to the rain forests and is found widely distributed over the peninsula. Its current known distribution extends from Indonesia, through Peninsular Malaysia, and up to Thailand. It is also a highly damaging species and can cause severe damage in unprotected situations. The distribution of Taxon A in cultivated areas only and Taxon B in cultivated as well as forested areas illustrates the need for detailed host record and distribution studies for other pest species in tropical Asia. Such knowledge is vital to the success of area wide control programs such as SIT.

The major pest species in the Philippines is now known to be an undescribed species in the *dorsalis* complex (Philippine B) and *B. occipitalis*, not *B. dorsalis*, is the major pest of mango.[5]

III. CURRENT CONTROL METHODS AND STRATEGIES

The various control methods, their use and applicability in Peninsular Malaysia have been reviewed by Vijaysegaran.[13] These include early harvesting at the green stage for papaya, banana and sapota; and fruit bagging or wrapping for carambola, mango, jackfruit, and cempedak. Protein baits formulated from locally available materials and applied as low volume spot sprays have also provided excellent control of flies infesting carambola.[14] Many of these methods are also used in other countries of tropical Asia. These methods center primarily around the population suppression in individual fruit orchards. Cover sprays of insecticides are commonly used in India[15] and the Philippines,[16] while protein bait sprays are used to a limited extent in Thailand[17] and Malaysia.[14,18] Wrapping of individual fruits in paper bags to prevent oviposition appears to be a popular and successful method unique to tropical Asia. This method also produces high quality fruit but its use over extensive areas is limited by availability of skilled labor.

Area wide population suppression programs using SIT have been reported from Thailand[19] and the Philippines,[20] and SIT and other means (male attractants) from Taiwan.[2] The actual success of these control programs is unclear. The attempt in the Philippines appears to be a short series of trials on mango which was then not continued. In Thailand, SIT trials at Antkhang in the highlands of Chaing Mai in the north, apparently encountered problems when release of sterile flies to control *B. dorsalis* and *B. correcta* were complicated by the resurgence of another species, *B. zonata*.[19] It is also suspected that flies in culture may have been contaminated by another nonpest species, *B. tuberculata*.[4] In Taiwan, SIT was used on a large scale beginning in 1975 and continued until 1984. The SIT program was then terminated and replaced with a male annihilation strategy using methyl eugenol. Significant reductions in damage to commercial citrus and mango was reported.[2]

IV. AREA WIDE CONTROL PROGRAMS AND THE SIT

The increasing growth and organization of horticultural industries in tropical Asia is evident and its continued success is largely dependant on sound fruit fly control measures. High quality fruits are required by importing countries in European markets. Countries like Japan that are a lucrative market, also have rigid quarantine regulations that often deny the entry of many tropical Asian fruits. It is for these reasons that area wide fruit fly control programs are becoming more important to many countries of tropical Asia.

There is growing interest in the use of SIT in tropical Asian countries, because of its previous application and use in other parts of the world. However, there are several situations and problems that are unique to the region. These are: (1) any method including SIT to be used for area wide control would have to meet the following criteria: (a) first, it would have to be environmentally acceptable, i.e. without deleterious side effects to non-target organisms, and (b) secondly, it would have to be cost effective and economically viable to implement; and (2) the objectives of area wide control would be twofold: (a) first, to achieve *population suppression* over large areas, and (b) then to attempt *eradication* and maintain quarantine to establish fly free zones.

A.   POPULATION SUPPRESSION

Suppression of fruit fly populations in individual orchards using insecticide cover sprays or protein bait sprays is an important strategy in enabling fruit production in many tropical Asian countries. These techniques are broad spectrum and control all species of infesting flies. Population suppression over large areas would undoubtedly provide better control and also benefit a large number of growers. A major consideration would be the cost, and if it can be shown that SIT is cost effective then there is no reason why continued releases of sterile flies should not be tested as an area wide control technique. Hooper[21] commented along similar lines.

An important consideration, however, would be the local species and their host ranges. Unlike previous programs in the past where SIT has been applied to a single species usually introduced to that region, the multiplicity of species in the Asian region presents a totally new challenge.

B.   ERADICATION

Apart from the success of the project in Japan,[22] eradication of a fruit fly species by SIT or other means has not been reported in other Asian countries. Any attempt at eradication in the future would have to take several factors into consideration. Taking peninsular Malaysia as an example for the region, several problems can be envisioned in an eradication based strategy. Taxon A and Taxon B, two of the major pest species, have 17 hosts in common (7 cultivated and 10 wild). A SIT program would have to consider the logistics of dealing with two major species with several common hosts instead of one. The widespread presence of the *B. cucurbitae*, *B. latifrons*, *B. tau*, and *B. umbrosa* would also require an eradication program to be conducted for eradication to have any significance and benefit from the viewpoint of quarantine and export related trade in fruits and vegetables.

For species like Taxon B whose distribution extends from cultivated areas through to rain forests, it would be difficult to identify areas that could provide some sort of geographic or ecological isolation for eradication to be achieved.

Post-eradication quarantine to prevent reinfestation is probably the biggest obstacle to any eradication attempt in the southeast Asian region. Reinfestation of an eradicated area could arise first from gravid flies migrating in from surrounding areas. Given the diversity of the natural vegetation in the region, this would be very difficult to control.

Invasion by exotic fruit fly species into the eradicated area from neighboring countries is also another very important consideration. The possibility of an exotic pest species invading the vacant niche (host plants) resulting from an eradication program is a real possibility for Thailand, Malaysia, Indonesia, the Philippines and other countries in the region. Unlike protein bait sprays or insecticide cover sprays that would control several species of fruit flies, any SIT related program would be species specific. Eradication of the target species would leave the area open with no protection against other fruit fly species.

Current survey and taxonomic work[5] shows that each country has its own pest species that may not be found in neighboring countries. For example, both *B. correcta* and *B. zonata* have a propensity to infest several fruit types of economic importance in Malaysia. Both species are present in neighboring Thailand but to date have not been reported from Peninsular Malaysia. The Philippine Islands and the Indonesian Archipelagos also possess several pest species that could easily colonize suitable economic hosts in other countries in the region. Given the extensive coastline, the movement of people and produce, and the geographic proximity of all these southeast Asian countries to one another, fruit flies would easily drift to new areas. It would need an enormous effort to enforce and maintain quarantine. Unless the quarantine related issue can be solved, it is unlikely that eradication could be considered a viable strategy to adopt.

The multiplicity of economic species in the region also presents the need for efficient rearing techniques. Many species have not been reared on artificial media before. Experience in Malaysia has shown that some species adapt well to laboratory culture while others continue to be plagued by low egg production and hatchability.[23,24] The current data bank on medfly, oriental fruit fly and melon fly though vast, may not be entirely applicable to the many new species in the region. Quality control procedures, handling and transport, release techniques, etc. all require further research in the respective countries.

V. CONCLUSIONS

The rapid expansion and economic importance of horticultural industries in the Southeast Asian region is dependent on sound fruit fly control programs. Given the current shift from traditional backyard cultivation to large scale commercial fruit orchards, it is only logical to consider implementation of area wide control programs that would be aimed either at suppressing pest fly populations over large areas or to try to achieve eradication or fly free zones.

Area wide population suppression programs using a combination of protein bait sprays, sanitation, male trapping and selective insecticide cover sprays are being used to varying degrees in the different countries in the region. The use of SIT as a population control tactic in such a program deserves consideration. There is no reason why it could not be tested as a component of a population suppression program, providing it is cost effective. The use of SIT for eradication is a formidable task. In view of the presence of many previously undescribed pest species in the Southeast Asian region, the geographical proximity of these countries to one another and the difficult task of enforcing quarantine to keep out exotic species, it is perhaps best if the concept of eradication encompassed a number of countries rather than individual countries. This would incorporate harmonization in control efforts and plant quarantine regulations and procedures that are vital to the success of such a huge and ambitious venture.

REFERENCES

1. Singh, R. B., Fruit flies in the Asia Pacific region, in *Proc. 1st. Int. Symposium on Fruit Flies in the Tropics,* Kuala Lumpur, Malaysia, 1988, Vijaysegaran, S. and Ibrahim, A. G., Eds., 1991.

2. Cheng, C. C. and Lee, W. Y., Fruit flies in Taiwan, in *Proc. 1st. Int. Symposium on Fruit Flies in the Tropics,* Kuala Lumpur, Malaysia, 1988, Vijaysegaran, S. and Ibrahim, A. G., Eds., 1991, 152.

3. White, I. M. and Elson-Harris, M. M., *Fruit Flies of Economic Significance: Their Identification and Bionomics,* C.A.B. International, Wallingford, U.K., 1992, 601 pp.

4. Drew, R. A. I., personal communication, 1992.

5. Drew, R. A. I. and Hancock, D. L., The *Bactrocera dorsalis* complex of fruit flies (Diptera: Tephritidae), *Asia Bull. Ent. Res.* (In Press).

6. Drew, R. A. I., Taxonomic studies on oriental fruit fly, in *Proc. 1st. Int. Symposium on Fruit Flies in the Tropics,* Kuala Lumpur, Malaysia, 1988, Vijaysegaran, S. and Ibrahim, A. G., Eds., 1991.

7. Elson-Harris, M., Studies in larval taxonomy of tropical fruit flies, in *Proc. 1st. Int. Symposium on Fruit Flies in the Tropics,* Kuala Lumpur, Malaysia, 1988, Vijaysegaran, S. and Ibrahim, A. G., Eds., 1991.

8. **Ooi, C. S.,** Genetic variation in populations of two sympatric taxa in the Dacus dorsalis complex and their relative infestation levels in various host fruits., in *Proc. 1st. Int. Symposium on Fruit Flies in the Tropics,* Kuala Lumpur, Malaysia, 1988, Vijaysegaran, S. and Ibrahim, A. G., Eds., 1991, 30.

9. **Vijaysegaran, S.,** The occurrence of oriental fruit fly on starfruit in Serdang and the status of its parasitoids, *Journal of Plant Protection in the Tropics,* 1, 92, 1983.

10. **Hancock, D. L.,** Identification of the Dacus species (Diptera: Tephritidae) present in Suriname, F.A.0. Project TCP/RLA/8858, Technical Report No. 1, 1989, 3 pp.

11. **Hancock, D. L.,** The carambola fruit fly, Dacus sp. near dorsalis (Diptera: Tephritidae) in Suriname, F.A.O. Project TCP/RLA/8858, Technical Report No. 2, 1990, 17 pp.

12. **Ferrar, P.,** Scent of doom for Malaysian fruit flies, in *Partners in Research for Development, Australian Centre for International Agricultural Research, No. 3,* 1990, 2.

13. **Vijaysegaran, S.,** Management of fruit flies, in *Integrated Pest Management in Malaysia,* Lee, B. S., Loke, W. H., and Heong, K. L., Eds., Malaysian Plant Protection Society, 1985, 231

14. **Vijaysegaran, S.,** An improved technique for fruit fly control in carambola cultivation using spot sprays of protein baits, National Seminar on Carambola, 18-19 July, Kuala Lumpur, Malaysia, 1989.

15. **Agrawal, N. and Mathur, Y. K.,** The fruit fly problem , associated with cultivated crops in India and its control, in *Proc. 1st. Int. Symposium on Fruit Flies in the Tropics,* Kuala Lumpur, Malaysia, 1988, Vijaysegaran, S. and Ibrahim, A. G., Eds., 1991, 191.

16. **Rejesus, R. S., Baltazar, C. R., and Manoto, E. C.,** Fruit flies in the Philippines: Current status and future prospects, in *Proc. 1st. Int. Symposium on Fruit Flies in the Tropics,* Kuala Lumpur, Malaysia, 1988, Vijaysegaran, S. and Ibrahim, A. G., Eds., 1991, 108.

17. **Meksongsee, B., Liewvanich, A., and Jirasuratana, M.,** Fruit flies in Thailand, in *Proc. 1st. Int. Symposium on Fruit Flies in the Tropics,* Kuala Lumpur, Malaysia, 1988, Vijaysegaran, S. and Ibrahim, A. G., Eds., 1991, 83.

18. **Vijaysegaran, S.,** The current situation on fruit flies in Peninsular Malaysia, in *Proc. 1st. Int. Symposium on Fruit Flies in the Tropics,* Kuala Lumpur, Malaysia, 1988, Vijaysegaran, S. and Ibrahim, A. G., Eds., 1991.

19. **Suntantawong, W.,** Control of fruit flies Dacus dorsalis Hendel and Dacus correctus Bezzi by sterile insect technique at Ang Khang, Chaing Mai, Report of the Office of Atomic Energy for Peace (in Thai), 1988.

20. **Manoto, F. C., Blanco, S. S., Resilva, S. S., Baturi, E. S., and Cordero, A. D.,** Control of oriental fruit fly by the sterile male release technique: studies ᴏₙ population dynamics, fruit infestation sequence and dispersal of Dacus dorsalis Hendel, Term. Rept. PAEC, Diliman, Quezon City, 1980, 40 pp.

21. **Hooper, G. H. S.,** Fruit fly control strategies and their implementation in the tropics.,in *Proc. 1st. Int. Symposium on Fruit Flies in the Tropics,* Kuala Lumpur, Malaysia, 1988, Vijaysegaran, S. and Ibrahim, A. G., Eds., 1991, 30.

22. **Kawasaki, K.,** Eradication of fruit flies in Japan, in *Proc. Int. Symposium on Biology and Control of Fruit Flies,* Okinawa, Japan, 1991, 22.

23. **Vijaysegaran, S.,** unpublished data.

24. **Suleiman,** personal communication, 1992.

Chapter 12

BIOCLIMATIC EFFECTS ON THE DISTRIBUTION OF THE MEDITERRANEAN FRUIT FLY (DIPTERA: TEPHRITIDAE) IN THE MAGHREB

E. J. Buyckx

I. INTRODUCTION

The Mediterranean fruit fly (medfly), *Ceratitis capitata* (Wiedemann) is the only polyphagous fruit fly found in Northwest Africa. In 1988, at the request of the four infested countries, Libya, Tunisia, Algeria, and Morocco, the International Atomic Energy Agency (IAEA) established a project entitled "Survey on the Extent of Medfly Infestations in North Africa" as a preparatory step to determine the feasibility of a medfly control and eradication program in the Maghreb.

In modern times, the word MAGHREB - in Arabic "MARHRIB" or "MAGHRIB" meaning the West - has been used to designate the countries of Northwest Africa lying between the Sahara and the Mediterranean Sea, comprising Algeria, Morocco, and Tunisia. In this paper, the Maghreb is referred to as the area west of Egypt and stretching to the Atlantic Ocean. Ancient Arab geographers subdivided it into three large regions: "IFRIQIYYA" (Tripolitania and Tunisia), the "MAGHRIB AL-AWSAT" (Algeria), and the "MARHRIB AL-AQSA" (Morocco).

The medfly was first reported in 1855 from Tunisia[1] and about the same time from Algiers, Algeria.[2] It was later reported from Morocco and the Libyan Arab Jamahiriya. So far, it has not been reported from Mauritania. One important factor in estimating the resources required to implement an area-wide control/eradication program is the distribution of the pest in the whole region. At present, this is not well known. With the assistance of the IAEA, the interested countries have initiated medfly trapping programs and fruit infestation studies to determine the limits of infested areas and the annual population fluctuations of the pest. Hopefully, these efforts will result in a distribution map of the medfly in the Maghreb. However, the region is very large, totalling 4,758,340 km² for Morocco including the northern provinces only, most of Algeria and Libya lying in the Sahara. It may take a few years before such a map is available due to limited resources available for extensive trapping programs. For a general approximation of the infested range of the medfly, this paper is an attempt to estimate the areas infested in the Maghreb on the basis of the bioclimatology of the region and of what is known about host plant occurrence.

A. RAINFALL IN NORTHWEST AFRICA

Most of the land in the Maghreb countries is desert, and only the coastal plains and mountain masses lying in the path of humid winds coming from the Atlantic Ocean receive enough water for rain-fed and irrigated agriculture. In Mediterranean Africa, the gradient of change in rainfall is in the range of 10% for 100 m change in sea level and one degree latitude or about 0.25 to 1.0 mm per kilometer.[3] Rainfall nearly doubles for each increase of 1,000 m in altitude.

The distribution of rainfall plays a predominant role in the occurrence and fruiting of host plants of the medfly in the Maghreb, due to the presence of the Atlantic Ocean on the west, the Mediterranean Sea on the north, and the Sahara Desert on the east and south. The region exhibits a Mediterranean climate with the rainy season usually extending from September/October to April/May. This period is usually followed by a sunny and dry summer. In the Mediterranean Basin, two rainfall patterns are observed. A bimodal pattern of rainfall prevails in the western part, while a monomodal pattern occurs in the eastern part. Most of the rainfall occurs in autumn in the coastal and lowlands areas, while the highland and continental areas receive a rainy period in autumn and a second one in spring. In general, bimodal patterns are favorable for permanent herbaceous and woody species and monomodal patterns are favorable for annual plants with a short development cycle.[4]

Rainfall distribution is quite variable for a given year.[5] This variability is inversely related to the mean and is an extremely important feature of arid zones.[6] Natural vegetation and rain-fed crops are negatively affected by this variability. The main characteristic of the climate of the Maghreb is its great diversity, resulting from the small width of climatic zones under the combined influence of the sea, altitude, and latitude.[5]

The Atlas Mountains extend from north of Agadir, Morocco in a west southwest to east northeast direction terminating in central Tunisia (Map 1). The major peak, Jebel Toubkal (4,167 m) in the High Atlas in Morocco. The Saharian Atlas in Algeria has peaks of up to 2,235 m in the Ksour Mountains, and the Aurès has the Jebel Chelia (2,328 m). The Tebessa range reaches 1,712 m near the Tunisian border, with the occurrence of Jebel Chambi (1,544 m) in central Tunisia, and the Medjerda mounts (Jebel Ouergha, 912 m) in northern Tunisia.

South of the High Atlas range in Morocco and more or less parallel to it, forming the southern limit of the Sous Valley, rises the Anti-Atlas range. It is linked to the High Atlas at its center by its highest point, the Jebel Siroua (3,304 m). The Middle Atlas to the north of and parallel to the High Atlas but not as high, rises progressively, in a northeasterly direction, to above 3,000 m with the peak of Jebel Bou Naceur (3,340 m) dominating the upper valley of the Moulouya river.

The Rif range consists of a series of coastal mountains stretching eastward from the Strait of Gibraltar to the Moulouya valley, forming a bend towards the Middle Atlas from which it is separated by the Taza pass (585 m). The Jebel Tidighine, reaching 2,452 m, is its highest peak.

The western slopes of the Rif, the Middle Atlas, and the High Atlas form a barrier forcing the winds from the Atlantic to rise and lose their moisture. To illustrate this phenomenon, annual rainfall at the Jebel Outka in the Rif reaches 1,739 mm and 1,467 mm at Bab-bou-Idir in the Middle Atlas, while in the Moulouya Basin, it drops below 200 mm.[5] As a result, the part of Morocco between the Atlantic coast and the mountain ranges receives more rain than any other region of the Maghreb; no other part of North Africa can be compared to it. On the other hand, leeward areas beyond the eastern slopes lie in a rain shadow with much lower annual rainfall (e.g. the Moulouya valley). This is partly true for the Anti-Atlas and south of the Saharan Atlas as well where the mean annual rainfall drops suddenly below 100 mm.

In Algeria, the coastal Tell mountain ranges, (Atlas Tell) with a peak of 2,308 m in the mountain mass of Jurjura, delimit with the Saharan Atlas a series of tablelands called the High Plains. The altitude in the western part decreases progressively from 1,200 m near the Moroccan border to 400 m in the Hodna depression, while in the eastern part, it varies between 900 m and 1,200 m. The bioclimatology of this region was studied by Le Houérou et al.[7]

In Libya, lowlands and mountain ranges of comparatively modest elevation occur in northern Tripolitania and northern Cyrenaica. Much of the interior is comprised of plains above 500 m in altitude, diversified by escarpments and volcanic outcrops. The mountains of northern Libya are of less complex structure than the Atlas mountains and occur in two separate ranges, one in Tripolitania and the other in Cyrenaica. The Tripolitanian range or Jebel Nefusa, extending from the Tunisian border to Misratah, consists of two series of ridges, the highest of which reaches 800 m, with a peak of 981 m, and includes from west to east Jebel Jefren and Jebel Garian. In Cyrenaica, the Jebel el Akhdar (882 m), similar to the highlands of Tripolitania is, on the average, slightly higher.

The climate is continuous from the eastern part of Morocco beyond the Rif to Tunisia. Although quite arid in the Moulouya Basin (200 to 400 mm), it becomes more humid moving to Tunisia (maximum 1,607 mm at Aïn Draham, Kroumiria, at 800 m), but becomes drier from north to south. The isohyets run nearly parallel to the Mediterranean coast with some bends around mountainous areas. For instance in Algeria, the coastal area is narrow, a maximum width of only 50 km in the plains around the Sebhra of Oran and in the Seybouse river valley, with a mean annual rainfall of less

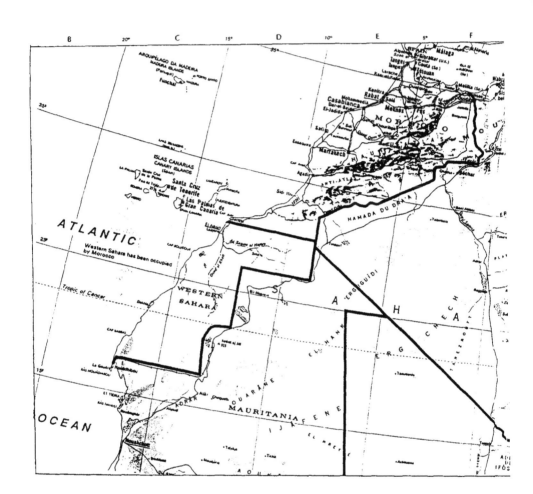

MAP 1 OF THE MAGHREB. Reproduced from the map for North Western Africa, page 124 of Reader's Digest Atlas of the World, with kind permission from Rand McNally & Co.

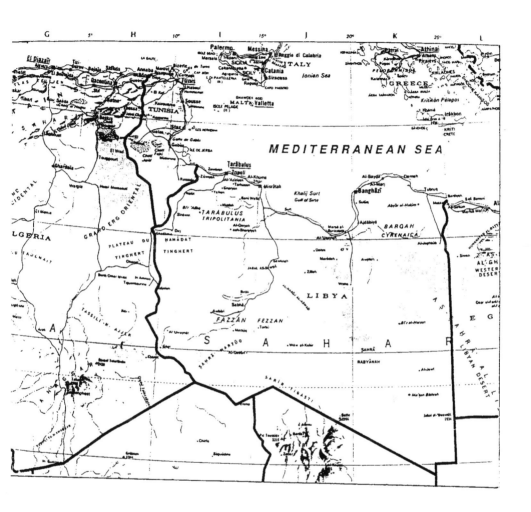

than 400 mm in the low plains of the Oran region to above 1,000 mm in the coastal area of Constantine. As a result, this region is divided into a western littoral climate that is relatively arid and of the steppe-type in the wilaya of Oran, and an eastern littoral climate that is relatively wet. Seltzer[5] draws the limit between the two zones to the lower part of the oued Isser. Eastward of this limit, annual rainfall is about 800 mm, and above but westward, it is below 800 mm.

The Atlas Tell cause winds coming from the sea to rise, thus producing condensation and rainfall. This is the wettest part of Algeria; most of the rainfall occurs on those mountain masses that are highest and nearest to the coast, more so in the eastern part than in the western part. The least rainfall is leeward of these mountains. The Tell High Plains receive an annual rainfall above 400 mm while in the Steppe High Plains, it is below 400 mm.[7]

The Saharan Atlas Mountains differ from the High Plains by the increased rainfall, although it is relatively low due to its distance from the sea. The peaks of the Aurès probably receive almost 800 mm of rainfall annually. Beyond the Saharan Atlas, the transition to the Saharan climate occurs suddenly. Rainfall is low interspersed with long periods of drought.

The extension of Libya south of the 32° latitude has a relatively drier climate compared to northern Algeria and Tunisia lying south of 37° latitude. A maritime climate, in which rainfall occurs between October and March, prevails only in a narrow coastal belt about 5 to 10 kilometers wide. Agriculture is located mainly in that zone and most of the fruit is produced there. As is the case for northern Algeria and Tunisia, the amount of rain (100 mm to 400 mm average) increases eastward from the Tunisian border to Tripoli, then falls below 100 mm to the east side of the Gulf of Sirte. In the Cyrenaican highlands, it increases from over 200 mm up to over 600 mm in the Jebel el Akhdar. South of the coastal strip, the major part of the country (94%), the desertic zone, has less than 100 mm average annual rainfall, including the true desert (58% of the country) where precipitation is virtually zero.

B.   RELATIVE HUMIDITY
Relative humidity is quite high in the coastal plains of the Maghreb, especially in summer, with an annual average of 75-85% along the Atlantic coast, 65-75% on the Mediterranean coast, while on the continental steppe of the arid zones it varies from 60-65% with a marked maximum in winter and a strong minimum in summer.[8] In the desertic zone between the 25 mm and 100 mm isohyets, the annual average ranges from 40-50%, dropping to almost 30% in summer. In the true desert, it varies from 25-40% with a minimum as low as 10-25%. For example in Libya, the annual average relative humidity ranges from 60-70% in the coastal zone to 50-60% in the

steppe zone, while in the highlands, it varies from 55-65% in the Jebel Akhdar and 50-55% in the Jebel Nefusa.[9]

C.   SUB-CLIMATES OF NORTHWEST AFRICA

The climatological and biological criteria for climates south of the Tropic of Cancer are summarized in two tables according to sub-climates.[10] A similar classification of the sub-climates occurring in the four Maghreb countries on the basis of the criteria established by Gounot and Le Houérou[8,11,12] is shown in Table 1. Starting from north to south, eight main zones may be distinguished. The first zone, hyperhumid, is the wettest, with a mean annual rainfall (P) above 1,200 mm. It consists in rather limited areas near the sea in Algeria and Tunisia, where forests predominate. In the second, the humid zone, where P varies between 800 and 1,200 mm, there are several wild host plants of the medfly and a wide variety of food and industrial crops. Fruit trees are grown around farms and villages. In the third, sub-humid zone, with P from 600 to 800 mm, mainly cereals, vegetables, leguminous, and fodder crops are produced.

In the fourth, the semi-arid zone, P ranges from 400 to 600 mm and most arable land is used for cereal production. Many fruit trees are grown, but olives and apricots become dominant and prickly pear cactus is common. There are fewer citrus orchards than in the preceding zone because irrigation is often required to obtain a good crop.

In the last two sub-climates, population densities are relatively high (60 to 100+ inhabitants/km). For example, in the Bizerte Governorate in northern Tunisia, with a mean annual rainfall of 600 to 800 mm (sub-humid with mild winters), the density is more than 300 inhabitants per km^2. Farming is prevalently on an individual basis and is dispersed over the whole arable land area. Villages are more commercial and administrative centers than farm concentrations. Wild host plants grow practically everywhere and fruit production (citrus, peaches, apricots, pears, apples, figs, medlar, quince, prunes, etc.) is widespread in orchards, fruit gardens, and backyards. However, vast stretches are devoted to cereals only.

The fifth or arid zone with P between 100 and 400 mm, a steppe of dwarf shrub and perennial grasses, extends over large tracts where wild host plants such as jujube and prickly pear occur. The 400 mm isohyet is the upper limit of steppe vegetation.[8] The high plateaus of Algeria and western Tunisia are covered with esparta grass or "alfa,"[7] while in east central Tunisia, the landscape is dominated by extensive olive groves with small orchards of apricots, almonds, pistachios, fig trees, vineyards, date palms, and pomegranates. Fig, mulberry, apricot, and bitter orange are often interplanted in olive groves. Other host fruit trees such as citrus, peaches, apples, pears, pomegranates, ornamentals, etc. are limited to back yards and public gardens.

Table 1

Ecoclimatic Classification of Northern Africa[8,11,12]

Sub-Climates	Rainfall mm (P)	$100P/PET_p$ (R)	Length of Rainy Season (RS) in "Humid" Months	Vegetation Types/Livestock	Land-Use Patterns
Eremitic	P<25	R≤1	RS = 0	Desert, no vegetation without underground water.	No rainfed land-use.
Lower hyperarid	25<P<100	a<R<3	RS = 0	Clustered - contracted vegetation, perennial grasses present. Camels, goats occasionally present.	Occasional long-range nomadism, no cultivation without irrigation. Oases (date palm, fruit trees, vegetables).
Upper hyperarid	50<P<100	3<R<6.5	RS = 0	Contracted vegetation, sparse dwarf shrub, argan tree, perennial grasses present, sheep, goats and camels, no cattle.	Nomadism, transhumance. No settled animal husbandry. No cultivation without irrigation. Oases (date palm, fruit trees, vegetables).
Arid (lower (middle	100<P<200 200<P<300	6.5<R<15 15<R<21	RS = 1 - 3	Steppe, dwarf shrub, perennial grasses locally dominant (alfa or	Transhumant and settled pastoralism, some retreat flood

Semi-arid	400<P<600	RS = 3 - 5	Sclerophyllic oaks, Aleppo pine, Phoenician juniper, shrubs. Cattle, sheep, goats, horses, donkeys; camels rare. Diversified rainfed cropping.	Pastoralism and barley farming. Rainfed cultivation of wheat, barley, fodder crops, sown pastures. Olives, stone and pome fruit, figs, grapes, citrus (irrigated).
Sub-humid	600<P<800	RS = 4 - 5	Woodland (sclerophyllous oaks), cropland. Cattle, sheep, goats, horses, donkeys. Diversified rainfed cropping	Pastoralism and sedentary livestock. farming barley, wheat, sugar beet, chick pea, maize; fodder crops and sown pastures possible. Olives, stone and pome fruit, citrus, figs, grapes.
Humid	800<P<1200	RS = 6 - 8	Woodland (deciduous and sclerophyllous oaks), open forest, cropland, perennial grasses. Cattle, goats, sheep.	Wheat, barley, maize, sugar beet, cotton, tobacco, rainfed rice. Sown pastures, fodder crops. Cork and timber production, meadows, stone and pome fruit.
Hyperhumid	1200<P	RS = 9 - 10	Timber and cork forest, meadows, woodlots (eucalyptus), cattle, goats.	Summer annual crops (Maize and sorghum), fodder crops, meat and milk, stone and pome fruit.

High plateaus are sparsely populated (about 10 to 30 inhabitants/sq. km). Most people live in villages and small towns, often stretched out along main roads. Agriculture is chiefly pastoral farming.

In North Africa, fruit can be produced under dry farming conditions with a mean annual rainfall in a bimodal pattern of 150 to 200 mm and quotients of 100 P/PET* of 10 to 15.[4] Prickly pear occurs extensively in rural areas. The argan tree grows over large areas at low densities near the southern tip of the Great Atlas, mainly in the Sous Valley, and in the Anti-Atlas. The rural population tends to concentrate in villages along oueds, springs, or wells. Many fruit trees, such as citrus, occur in back yards. Barley is still produced without irrigation while crops under irrigation are mostly vegetables and fodder. Fruit production in Libya and an important citrus industry in the Sous Valley in Morocco are possible under these conditions thanks to well irrigation.

The sixth zone, upper hyperarid, with a P between 50 and 100 mm, presents a diffuse natural vegetation of scrub and perennial grasses. The 100 mm isohyet is the upper climatic limit of the Saharan vegetation, hence the northern boundary of the Sahara desert.[8] Population density is very low (0 to 10 inhabitants/km²).

In the seventh zone, lower hyperarid, with less than 50 mm of rain per year, life is limited to very small areas with underground water resources such as oases and dry oueds. In oases, date palm dominates while many other fruits are grown under irrigation, such as citrus, peaches, pomegranates, figs, apricots, peppers, tomatoes, etc. The last zone, eremitic sub-climatic, is the true desert with P under 25 mm and no vegetation is present except where there is underground water.

D.   TEMPERATURE

In the Maghreb, mountains are subject to lengthy periods of snow cover. In Algeria, the high plateaus may have more than 50 days with frost, including late spring frosts. The seasonal distribution of temperature may be a limiting factor in plant development under Mediterranean mountain climates according to Le Houérou[4] who points out that a pause in plant growth is generally observed under temperate and Mediterranean climates during those periods in which the monthly mean of minimum temperatures during the coldest month (m)** is equal to or below 10°C. This mean minimum temperature corresponds to an average of daily minimums of 5.0 ± 1.5° C. For a better understanding of plant physiology in those areas, Le Houérou[4]

---

*   P = mean annual rainfall.
         PET = potential evapo-transpiration measured by the Penman method.

**   m = mean of minimum daily temperatures of the coldest month.

has proposed a second parameter M'* closely correlated to the evapo-transpiration of the coldest month and to the ability of plants to grow on warm sunny days when nights are cold with m dropping to 1 to 3°C. For example, in the Algerian high plains and similar regions in Libya, Morocco and Tunisia, during the coldest month, for m < -2°C, the mean daily maximum temperature M' is below 10°C; with m = 1 to 3°C, M' may range from 10 to 15°C; for m = 3 to 5°C, M' = 12 to 15°C; for m = 5 to 7°C, M' = 15 to 20°C. According to Le Houérou[9] for m between 3° and 5°C, characteristic of temperate winters with 10 to 20 days of light frost, plant development slows down in December-January. Values of m between 1° and 3°C characterize cool winters having 20 to 40 days of frost with growth of local plants reduced from mid-November to mid-February. In Mediterranean areas, periods favorable to plant growth correspond to a combination of total annual rainfall about 1/3rd of PET with mean minimum temperatures higher than 5°C.[9] The following bioclimatic features of some values of m have been described by Le Houérou.[4,8] The isotherm of m = 5°C for January delimits areas for which the physiological winter pause is noticeable; it is the lower temperature limit for plants susceptible to cold, such as citrus, Opuntia cactus, and phyllodial acacia trees, hence for the commercial production of citrus.

The isotherm for m = 3°C delimits areas with 10 to 20 days of frost under shelter, in which plant species related to winter cold appear; it draws the line between high and low plains, the altitude being generally under 600 m, and it is the absolute temperature limit of species susceptible to cold, such as citrus, opuntia, and *Acacia* spp. Such areas are favorable for growing fruit trees requiring low temperatures for fructification, such as apple, pear, cherry, pistachio, etc. The isotherm of m = 1°C is the lowest temperature limit of the olive tree and the date palm. The altitude is generally higher than 600 m but below 1,000 m in the continental zone. Areas having cold winters with m between -1° and 1°C are located at an altitude above 1,000 m in the continental zone but much higher in the oceanic zone. A value of m = -2°C (above 1,500 to 2,000 m according to latitude) is the lower limit of cedar and fir trees, while the upper limit of trees in Mediterranean Africa varies from 2,500 to 3,500 m.

Gounot and Le Houérou[13] have successfully sub-divided the northern half of Tunisia into the humid, sub-humid, semi-arid, and arid sub-climates and their variants, incorporating the mean temperatures of the winter months, thus providing a useful tool for planning agricultural development. Projecting the isotherms for m and M', on their bioclimatic map of northern Tunisia, these authors[13] determined the winter variants - cool, temperate, mild, warm - for each sub-climate of that part of the country, as well as for M' = 10°C and m = 3°C according to altitude. As might be expected, the areas with

---

*   M' = mean of maximum daily temperatures of the coldest month.

mild and warm winters are to be found along the Mediterranean Sea. The farther inland one proceeds, the cooler the winter becomes. Thus, in the Maghreb countries not only the climate of internal regions gets drier but cool winters become colder and longer with altitude.

During summer, maximum temperatures in arid, hyperarid, and desertic sub-climates may exceed 50°C like at Taroudant in the Sous Valley. According to Le Houérou,[9] in Libya the mean maximum temperature M* in the coastal strip reaches 28-32°C in July-August, 32-38°C in the Tripolitanian hinterland and 38-42.5°C in the desert.

E.   OCCURRENCE OF HOST PLANTS

Liquido et al.[14] have listed those plant species reported in the world as hosts of *C. capitata*. Of a total of 353 species, only 199 have field and/or laboratory infestation data. Of these, 64 wild and cultivated host plants are known for the Maghreb (Table 2).   For ten species, the presence or development of eggs or larvae in fruit collected in the field or infested in the laboratory has not been reported. From the fruit of the other species, adults were obtained in at least one of the four countries. However, few studies have been published.

Among the host plants, three deserve particular attention, because in the southern part of the region, dates and prickly pear, and in Morocco only the argan tree, may be of great importance for the survival of medfly populations.

1.   Date Palm

In the Maghreb countries, only date palms (*Phoenix dactylifera* L., *Palmaceae*) growing in the continental hyperarid zone with less than 100 mm mean annual rainfall and values of m above 1°C and below 5°C, produce quality fruit suitable for export.[8] Dates mature from September to December up to an altitude of about 1,000 m,[8,15] while only in the mountains of the southern Sahara (Tibesti, Chad) are fresh dates presumably produced up to 1,500 m.[15]

Medfly infestations in dates have been observed in several countries.[16-20] In southern Algeria, infested dates were reported from the oases of El Golea[21] and Ghardaia. In Libya, Blak et al.[22] collected flies from ripe dates of two local varieties, Karkabi and Hallaawi, in September. Medfly larvae have also been found in dates in Morocco.[23]   Although, medflies were observed resting on leaves and dates of palms surrounding an unsprayed orchard of guavas, oranges, mangoes, and grapes, Hendrichs and Hendrichs[24] considered date palm as a non-host in Egypt.   In the FAO paper[15] on date

---

*   M = mean of maximum daily temperatures of the hottest month.

Table 2
Host plants of *Ceratitis capitata* (Wiedemann) in the Maghreb

Family Botanical Name	Common Name	Country*
Apocynaccae		
Carissa macrocarpa (Ecklon) A.DC.	Natal plum	T
(= C. grandiflora (E.H. Mey.) A. DC.		
Asclepidiaceae		
Calotropis procera Dryander ex Aiton.f.	Aak, Sodom apple	A, L, M**
Cactaceae		
Opuntia dillenii (Ker.-Gawl.) Haw.	Dillen prickly pear	A, L, T**
Opuntia ficus-indica (L.) Mill. (f. inermis Weber)	Spineless cactus,	A, L, M, T
(f. amycloea (Ten.) Weber	Indian fig	
Opuntia vulgaris Miller	Prickly pear,	A, L, M, T
	Barbary fig	
Capparaceae		
Capparis spinosa L.	Caper	A, L, M, T
Caricaceae		
Carica papaya L.	Papaya, pawpaw	L, M
Ebenaceae		
Diospyros ebenum J. Konig ex Retz	Black sapote	L, T
(D. ebenaster Retz.)		
Diospyros kaki L.f.	Oriental persimmon,	A, L, M, T
	Japanese persimmon	
Ericaceae		
Arbutus unedo L.	Strawberry tree	A, M**
Arbutus pavarii Pamp.	Strawberry tree	L**
Flacourtiaceae		
Dovyalis caffra (Hook. f. & Harv.)		
Warb. (= Aberia caffra Hook f. & Harv.)	Kei-apple	A, L, M, T
Juglandaceae		
Juglans regia L.	Walnut	A, M
Lauraceae		
Persea americana Mill.	Avocado	M, T**

* A = Algeria, L = Libya, M = Morocco, T = Tunisia.
** Plants listed only, no field and/or laboratory infestation data reported.

Family Botanical Name	Common Name	Country
Moraceae		
Ficus carica L.	Fig tree	A, L, M, T
Morus alba L.	White mulberry	A, L, M, T
Morus nigra L.	Black mulberry	A, L, M, T
Morus rubra L.	Red mulberry	L
Myrtaceae		
Psidium guajava L.	Common guava	L, M
Psidium littorale Raddi (P. cattleianum Sab.)	Strawberry guava	L
Rhamnaceae		
Ziziphus jujuba Mill. (Z. sativa)	Common jujube, Chinese date	L, M, T
Ziziphus lotus L. (L.) Lenn.	Jujube	A, L, M, T
Ziziphus Spina-Christi Wild.	Jujube	A, L, M, T
Palmaceae		
Phoenix dactylifera L.	Date palm	A, L, M, T
Punicaceae		
Punica granatum L.	Pomegranate	A, L, M, T
Rosaceae		
Crataegus azarolus L.	Azarole hawthorn	A, L, M, T
Crataegus sp.	Hawthorn	A, M, T
Cydonia oblonga Mill. (C. vulgaris Pers.)	Quince	A, L, M, T
Eriobotrya japonica (Thunb.) Lindl.	Japanese medlar, loquat	A, L, M, T
Malus domestica Boekh.	Apple	A, L, M, T
Mespilus germanica L.	Medlar	L, T
Prunus armeniaca L.	Apricot	A, L, M, T
Prunus avium (L.) L	Sweet cherry	A, M, T**
Prunus cerasus L.	Sour cherry	A, M**
Prunus domestica L.	Common plum	A, L, M, T
Prunus persica (L.) Batsch	Peach	A, L, M, T
Prunus persica (L.) Batsch cv. nectarina	Nectarine	A, L, M,T
Prunus salicina Lindl.xP. cerasifera J.F. Ehrh.	Methley plum	L
Pyrus communis L.	Pear	A, L, M, T
Rosa spp.	Rose tree	T
Rutaceae		
Casimiroa edulis Llave & Lex.	White sapote	T
Citrus aurantiifolia (Christm.) Swingle	Lime, sour lime	A, L, M, T
Citrus aurantium L.	Sour orange, bigarade	A, L, M, T
Citrus limon L. Burm.f.	Lemon, sour lemon	A, L, M, T
Citrus maxima (Burm.) Merrill (C. grandis)	Pomelo, shaddock	A, L, M, T
Citrus medica L.	Citron, cedrat	A, L, M, T
Citrus paradisi Macfady	Grapefruit	A, L, M, T

Family Botanical Name	Common Name	Country
Rutaceae (Cont'd)		
Citrus reticulata Blanco	Satsuma, mandarin, clementine, tangerine, orange	A, L, M, T
Citrus sinensis (L.) Osbeck	Navel and Valencia oranges	A, L, M, T
Fortunella japonica (Thunb.) Swingle	Kumquat	A, L, M, T
Fortunella margarita (Lour.) Swingle	Oval kumquat	L
Poncirus trifoliata (L.) Raf.	Trifoliate orange	L
Sapotaceae		
Argania spinosa (L.) Skeels (= A. sideroxylon)	Argan tree	M
Mastichodendron foetidissimum (Jacq.) Gronq. (= Sideroxylon mastichodendron Jacq.)	Ironwood	T**
Solanaceae		
Capsicum annuum L.	Red pepper, chili pepper, cayenne pepper	A, L, M, T
Capsicum frutescens L.	Tabasco pepper	A, L, M, T
Lycium europaeum L.	Box thorn	A, L, M, T
Lycium shawi (= L. arabicum Boiss.)	-	A, L, M, T**
Lycium subglobosum L.	Box thorn	T
Lycopersicon lycopersicum L. Karst. ex Farw. (=L. esculentum Mill.)	Common tomato	A, L, M, T
Solanum coagulans Schimp. ex Dun.	-	L, T**
Solanum melongena L.	Eggplant	L, T
Solanum sodomeum L.	Apple of Sodom	A, L, M, T
Vitaceae		
Vitis vinifera L.	Vinegrape, European grape	A, L, M, T

The information summarized in the above table has been collated from various papers and notes,[22,23,25-30] data kindly provided by H. N. Le Houérou,[31] and the lists of host plants presented by Ms. M. Bounfour, et al.[32] Botanical names and valid synonyms are those listed by Liquido et al.[14]

production and protection, the medfly is mentioned in the list of mites and insect pests of minor economic importance. It may be concluded that infestation of dates in the Maghreb is limited and of no economic importance. The importance of dates as a host in a hostile environment is related (1) to the time they reach a stage suitable for oviposition when other fruits have ripened and been harvested and (2) from their numbers. Even if infestations are slight and occasional, the large number of date palms present in an oasis could support quite a large medfly population for infesting other fruit reaching maturity in the area.

2.   Argan Tree

The argan tree, *Argania spinosa* (L.) Skeels (*Sapotaceae*), is found only in Morocco. It was widespread throughout the country in ancient times, but at present it exists only in extensive populations in the vicinity and south of Essaouira, around the southwestern tip of the Great Atlas and in the Sous and Draa Valleys. Some residual groups are found in the Beni Snassen mountain masses in the northeastern part of Morocco, on the slopes of the oued Grou near Rommani and Ezzhiliga, and south of Casablanca in the vicinity of El Jadida, Safi, Chemaia, and Chichaoua.[33] The isotherm for m = 3°C is the lower temperature limit for the argan tree at the southern edge of the great Atlas, while in the southern part of the Anti-Atlas, it is the isotherm of m = 3.8°C.[33]

The total area of the so-called argan forest extends over 700,000 ha according to Raf[33] but recent estimations give 600,000 ha. He divides it into two main zones. The first zone of dense vegetation, in which the argan tree reaches its full development, extends from Essaouira to Agadir and in the Sous Valley, up to 1,000 m on the slopes of the Great Atlas and south to Tiznit, altogether about 200,000 ha. In the second zone where the density drops to less than ten trees per hectare, it occurs along the fringe of the first zone, on the western slopes of the Great Atlas up to 1,500 m but mainly on the northern slopes of the Anti-Atlas, in the Draa Valley near Tindouf, and from there to the Atlantic Ocean.

The argan tree is xerophilous and thermophilous. It grows best in a mild climate with an annual rainfall between 120 to 500 mm.[33] It survives extended periods of hot weather (50.5°C), but it does not resist temperatures below 0°C.[23]

The argan tree has a high degree of polymorphism and, when growing under favorable conditions, resembles an olive tree, reaching a height of 8 to 10 m, as in the areas of Essaouira, Agadir, and Ait Melloul. Usually the vegetation remains green except when exposed to prolonged dry weather when it loses its leaves.[23] In a favorable environment, it bears fruit at various stages for most of the year, an important factor for the development of medfly populations.[23,34] However, most of the range of this species is not in favorable environments. There, the tree resembles a spiny bush or shrub and is practically the only plant growing over vast areas during the long dry season. Wandering herds of goats and camels feed on the leaves, twigs, and fruits. Goats are often seen climbing the trees. Under such conditions, trees have adapted to survive. They produce a limited amount of fruit that ripens at the end of the rainy season in March/April. Much of the fruit is eaten by domestic animals either on the tree or on the ground.

The argan fruit is a berry with a fleshy or a hard pericarp and a very hard seed and resembles an olive. Several medfly larvae may be found in one argan berry: Sacantanis[23] found infestations, east of Essaouira, varying from 0.07 larvae per fruit in April, reaching a peak of 20.6 in mid-July and declining to 5.1 by the end of November. In July, some fruits contained more than 40 larvae. A. Mazih[35] found more than 5,000 flies/trap/week in trimedlure-baited traps in the argan forest near Ait Melloul.

Sacantanis[23] believes that the argan "forest" does not provide year-round favorable conditions for the development of medfly populations because of the extremely dry weather. In the second zone, fruits are available for oviposition only during part of the year and medfly populations are able to persist only in spots where water is present in the ground, in oueds on the slopes of the Atlas Mountains, or in the vicinity of villages and irrigated areas. Although both Voegelé[34] and Sacantanis[23] stress the interest and importance of research on the life cycle and ecology of the medfly in the argan forest, no studies have been published.

3. Opuntia Cactus

A few species and several varieties of this cactus exist in the Maghreb, of which the most widespread are the spineless opuntia, *Opuntia ficus-indica* (L.) Mill., and the prickly pear, *O. vulgaris* L. Prickly pear is very common, is used as a fence to delimit property, to protect crops from wandering animals, for its fruit, as food for domestic animals, as a windbreak, and for erosion control. Spineless opuntia does not grow where the mean annual rainfall drops below 150 to 200 mm with values of m under 1.5° to 2°C.[8] This corresponds in western Tunisia and Algeria to a 900-1,000 m limit in altitude. However, in the Sous and Anti-Atlas, opuntia may be found up to 1,250 m when growing on steep slopes and narrow valleys facing southward.

Fruits ripen from August to November. They cannot be obtained in areas with an average daily relative humidity falling below 40% for more than a month.[29] When animals are allowed to eat the cladophylls or when opuntia are cut every two or three years for animal feed, they will not produce fruit.

The total area covered by opuntia in Tunisia comprising hedges around orchards, vineyards and farms, and plantations, was 100,000 to 150,000 ha in the early 1960s, of which 60,000 to 80,000 ha served as plantations for fruit or fodder, with about 35,000 ha in the steppe zone in the central part of the country.[29]

In a large part of the Maghreb, prickly pear, because of its fairly long fruiting season in late summer and autumn, serves as an intermediate host for medfly between summer and winter fruit. In Tunisia, Soria[29] found average infestation rates in fruit varying from 6.4 to 25%, with a peak at Menzel bou Zelfa in early September of 70%. He concluded that this fruit provides a last possibility to reproduce before the winter.

There are very few studies published on the development of the pest in prickly pears, although Bodenheimer[19] in Palestine, and Coupin[36] and Delanoue[37] in Tunisia, realizing the importance of this host for carrying over medfly populations, proposed to eliminate it from citrus-producing areas. Prickly pear infestation studies have been initiated in the four countries participating in the IAEA sponsored project.

II. DISCUSSION

A. CLIMATIC FACTORS AND THE MEDITERRANEAN FRUIT FLY

Cold weather is an important factor limiting the distribution of the medfly. According to Bodenheimer,[38] *C. capitata* is confined to regions where climate permits two generations a year and does not interrupt development for more than about 100 days. Avidov and Harpaz[20] found that females did not lay eggs at temperatures below 16°C and adults became inactive below 14°C. Flies cannot survive under 5°C for more than one to two weeks.[39]

In Marrakech, Morocco, during the winter of 1971, Harris[40] reared medflies from sour orange, Thompson orange, and artificial diet under controlled temperatures at 20° and 10°C, respectively. While he obtained many adults at the higher temperature, there were no survivors at 10°C.

Summarizing the limited data available on the development of *C. capitata* in relation to temperature, Meats[41] indicated lower development thresholds for eggs and larvae as 9.7°C, for pupae 13°C, and for adult activity 13.6°C. In Liguria and Sardinia, Italy, Delrio et al.[42] reported the lower limit for eggs was slightly under 10°C, while Crovetti et al.[43] indicated the lower limit for pupae was 11°C.

In extensive mountainous areas in Morocco, Algeria, and western Tunisia, winters are cold above 1,000 m in the north and central parts, and above 1,250 m in the southern part. With values of m = 3°C, (cold weather below 10°C for over 3 months), the medfly cannot survive.

In sub-climates where the P drops below 300 mm (arid and hyperarid) factors turn very unfavorable for the medfly. High temperatures combined with low relative humidities affect the adults, reducing their rate of reproduction[44] and egg development.[20,42] The upper development threshold for pupae is 35°C. However, when the relative humidity falls below 30%, less than one third survive.[41] Several authors have observed the destructive action on medfly populations of hot, dry southerly dusty wind (50°C and below 30% relative humidity) from the Sahara. It is referred to generally as the sirocco and called chergui in Morocco, ghibly in Libya, and khamsin in Egypt and the Near-East. The ghibly causes marked climatic variations. It occurs 30 to 90 days per year; the temperature rise above 40°C with relative air humidity dropping to 5-10% or less.[9]   According to Sacantanis,[23] the chergui is the greatest enemy of the medfly in the argan tree "forest." One or two days of the chergui kills all stages. He found millions of insects killed by the chergui in places sheltered from the wind, such as ditches. In Palestine, the occurrence of severe khamsin, like frost and inadequate rainfall, exerts a regulatory influence on medfly populations.   During khamsin weather, oviposition ceases and many females emerging during this period die without laying eggs.[20]

B.   AVAILABILITY OF HOST PLANTS

In the desert, host plants occur only in oases such as Ghardaia and El Golea in Algeria, separated by vast stretches of sand and/or stone. In the hyperarid sub-climate, wild host plants are scarce, limited to oueds, with the exception of the argan "forest" and prickly pear. Fruit trees are few, limited to villages located far apart in oases, along river beds, or in irrigated areas as is the case in Libya. In the arid sub-climate, the steppe harbors only widely scattered wild host plants and very localized cultivated host trees.

The areas with a semi-arid to sub-humid sub-climate (P between 400 and 800 mm) form the most favorable ecoclimatic zone for diversified rain-fed cropping, in particular horticulture. Most of the wild and cultivated fruit is available for oviposition most of the year.

In mountainous areas cold winters cause a scarcity of wild host plants, restricting the number of months during which fruit suitable for egg laying is available.

C.   DISTRIBUTION OF THE MEDITERRANEAN FRUIT FLY

Medfly can be found throughout the entire region wherever cultivated fruiting host plants occur. Although the medfly is quite mobile, movements appear to be restricted to a few hundred meters per week[45] when hosts are available and other conditions are favorable.   Soria and Cline[46] released laboratory reared flies labelled with ^{32}P among fruit trees near Tunis in November 1960 and June 1961. More than half of those recaptured were caught within a radius of 100 m and nearly half of the remainder within 200 m. The longest distance travelled was about 600 m by two flies one week after release. The number of flies recaptured was influenced by wind, by time since release, and by type of tree to which the traps were attached; peach and apricot were preferred. In a similar experiment in April 1962, Soria[47] observed a homogenous dispersal of labelled males from the site of release and at a maximum distance of 700 m within three days. Similar observations were reported by Hafez et al.[48] and Wakid and Shoukry[49] who released one day old, sterilized, laboratory reared flies into orchards in several semi-arid regions of Egypt in winter and summer. Movements of flies seemed to be influenced more by the distribution of hosts than by wind direction. During the releases in summer, flies appeared sluggish, possibly due to the heat, and were only caught at a maximum distance of 525 m.

Fletcher[50,51] concluded that most adults do not move great distances when hosts are available but, when fruit is unavailable, both immature and mature flies will disperse rapidly. Long distance flights, particularly over water, have been recorded. Movements of more than 20 km have been reported.[52] During experiments with the sterile insect technique on Procida Island, Bay of Naples, de Murtaset et al.[53,54] found that released flies dispersed more rapidly away from areas with no hosts than where hosts were present. Fertile females flew from the Italian mainland onto the islands of Procida and Ischia (2.7 km away) in August and started a new infestation.

Harris and Olalquiaga[55] monitored the occurrence and distribution of *C. capitata* in the desert coastal plain along the Pacific Ocean, near the Peru-Chilean border where temperatures range from 8 to 28°C with little or no rainfall. They found, in all frontier locations, a distinct pattern of increase and decline of fly populations during the year that were related to population trends in the Tacna Valley, Peru. Traps were distributed over distances of about 12 kilometers, 12 km south of Tacna. Flies were caught in locations without host plants but were found to be attracted to patches of ornamental trees that provided shelter in otherwise barren areas. Considering them as transient flies, Harris and Olalquiaga[55] conclude that in arid areas of Chile and Peru, *C. capitata* disperses over long distances to find favorable survival sites and that the synchronous dispersal pattern of males and females suggests it is an important behavioral surviving mechanism for medfly populations in such areas.

For the Maghreb, it seems likely that in the humid to semi-arid sub-climates, movements are limited when environmental conditions are favorable such as when several cultivated and wild host plants provide suitable fruit practically all year. However, in the arid and hyperarid sub-climates, movements over quite long distances, as observed by Harris and Olalquiaga,[55] are likely to occur as fruits suitable for ovipositing may be unavailable during certain times. Winds may not be expected to transport flies over long distances because of the high temperatures and low humidities.

Dispersal of medflies in the Maghreb is caused by the movement of fruit that is produced in the humid to semi-arid sub-climates in the coastal plains where widespread infestations develop every year. Modern roads allow fresh fruit to be transported throughout the countries, including to distant oases. Sacantanis[23] indicated that fruit transport and trade are the main dispersion factors. He observed the transport of Mediterranean fruit flies by car for a distance of 180 km to Marrakech, Morocco. It is obvious that in locations where conditions are unfavorable for *C. capitata* during part of the year due to cold winter months or lack of suitable fruit, medfly infestations may develop again in summer because of the importation of infested fruit.

III. CONCLUSIONS

Rainfall patterns and winter temperatures in the mountains in the Maghreb, determine areas unfavorable for the development of the medfly. Le Houérou[4] has estimated the areas included in the arid zones *sensu latu* and hyperarid zones in the Mediterranean region of North Africa. In Algeria, about 84% of the country is hyperarid and desert and thus not suitable habitat for medflies. In Libya, Morocco, and Tunisia, the proportion reaches 94%, 28.6%, and 38.5% respectively. This is about 81% of the total land area of the four countries including mountainous areas.

Areas have been estimated according to land use criteria from agricultural statistics contributed by the Maghreb countries.[56] Of a land area totalling 4,758,340 km^2, 3,775,170 or 80% is desert or high mountains; the total area of the Maghreb *sensu stricto* available for rain-fed agriculture is thus only about 980,000 km^2. This figure corresponds quite well with Le Houérou's estimates of 81%.

For the 980,000 km^2 where rain-fed agriculture is possible, the middle and lower arid areas that have less than 300 mm average annual rainfall are also considered unsuitable for the medfly. The land is extensive steppe with a scarcity of fruit due to very low host plant density, brief fruiting periods, and in the inland plains, cold winters and hot summers. An exception is a part of the argan "forest."

To delimit the area in Algeria, Morocco, and Tunisia in which a medfly trapping network should be established for determining the total infested surface, Buyckx and Vita[57] proposed the isohyet of 200 mm, the isotherm of a mean maximum annual temperature of 25°C, and the isohypse of 1.250 m. However, during follow-up visits to these countries and to Libya, it became clear that in areas receiving less than 300 mm of rain annually, the medfly can exist only in those places where irrigation makes fruit production possible. In sub-climates with mean annual precipitation above 300 mm, bimodal rainfall permits fruiting of wild host plants during two periods of the year. More intensive rain-fed cropping results in a greater density of cultivated hosts. On the coastal plains, mild winters and higher relative humidity are more favorable for the development and survival of the medfly. The isohyet of 300 mm, corresponding to the mean annual rainfall value separating the upper and middle levels of the arid sub-climates (Table 1, map 2), is proposed as the limit dividing Maghreb into two parts. Further, from the surface north of this isohyet, those mountainous areas with winter temperatures preventing the development of the medfly should be subtracted. They may be delimited by the isotherm of m = 3°C (Map 2). On the basis of these two criteria, the area of widespread infestation by *C. capitata* totals 220,000 km^2, with about 63,000 km^2 in Libya, 113,000 km^2 in Morocco, and 36,000 km^2 in Tunisia.

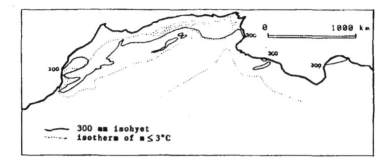

300 ㎜ isohyet
········ isotherm of ■ ≤ 3°C

Map 2. Bioclimatological map of the Maghreb showing rainfed areas with widespread Medfly infestations.

In examining the feasibility of eradicating the medfly from the Maghreb by the SIT, the above conclusions have a bearing on the strategy to be adopted and on the estimation of costs. It would make little sense to embrace an area-wide approach in releasing sterile flies as adopted in the MOSCAMED program in Central America to those areas with less than 300 mm mean annual rainfall and with values of m < 3°C. This approach would be suitable, however, for the large tracts of wild and cultivated hosts in 220,000 km² of rain-fed agricultural areas, to which should be added a part of the argan "forest" of possibly 2,000 km².

In arid and hyperarid zones, an integrated approach to eliminate the pest may be the most suitable solution. It should be based on the combined use of five control measures: strict fruit culling, particularly of sour orange and collection of fallen fruit, bait application, intensive trapping of flies (particularly females), efficient internal quarantine, and, where a residual population persists, local release of sterile flies. Bait could be applied by fruit tree growers using brushes of date palm leaves to spread the solution onto the host leaves. Although malathion is quite safe to use, a more acceptable substitute would be highly desirable.

The development of trapping devices and a powerful attractant for female flies would also be useful. El-Sayed et al.[58] evaluated trapping systems for the medfly in Assiut, Egypt during the period 1987-1991. He concluded that the McPhail trap (International Pheromones) with 300 ml of aqueous solution, of 9% nulure and 5% borax plus a trimedlure plug and a DDVP plug was the most efficient for capturing females. Such a trap caught nearly as many females as males. A simple method for disposing of, or neutralizing, infested and fallen fruit would be to fill black plastic garbage bags, seal them and leave them in the hot sun.

Of great importance would be the establishment and operation of internal quarantine to prevent reinfestation through movement of infested fruit from other parts of the country. The possibility of long distance movements of the pest needs to be considered as a source of reinfestation in eradication pilot trials and if medfly-free zones are planned.

Because detailed baseline information is lacking, it would be highly desirable to undertake pilot trials, not only for adapting the SIT to Maghreb conditions in well isolated areas, but also to carefully select localities within the arid and hyperarid zones for development of an integrated control program and to demonstrate that in this way medfly can be eliminated from such an environment.

ACKNOWLEDGEMENTS

I wish to thank Dr. W. Klassen, Acting Head, Insect and Pest Control Section, Joint FAO/IAEA Division of Nuclear Techniques in Food and Agriculture, for support and encouragement and for permission to reproduce the isotherm for m = 3°C and the 300 mm isohyet of the IAEA bioclimatic map 1992 for the Maghreb. The critical review of the manuscript and of the ecoclimatological component by Mr. H. N. Houérou from the Louis Emberger Center, CNRS, Montpellier, France, and the guidance in bioclimatology by Mr. G. F. Popov, Coordinator, Agrometeorological Group, Research and Technology Development Division, FAO, are gratefully acknowledged. Special thanks are also expressed to Ms. Lucia Kruzic for so ably processing the draft manuscript.

REFERENCES

1. La Gasca, F., La Mosca mediterrànea y la Uva de Almeria, Conferencia Accademia de San Ignacio, Almeria, 1925.

2. Vieira, R. M. S., A Mosca da Fruta (*Ceratitis capitata* Wied.) na Ilha da Madeira, Gremio dos Exportadores de Frutas da Ilha da Madeira, 1952.

3. Le Houérou, H. N., The desert and arid zones of northern Africa, in *Hot Deserts and Arid Shrub Lands, Ecosystems of the World*, 12B, Evenari, M., Noy-Meir, E., and Goodall, D.W., Eds., Elsevier, Amsterdam, 1986.

4. Le Houérou, H. N., Bioclimatologie comparative des zones arides s.l. de l'Afrique et de l'Amérique Latine, Terra Arida - Atelier Interrégional Afrique/Amérique Latine, MAB-UNESCO, Universidad de Chile, 7, Coquimbo, 1990, 25.

5. Seltzer, P., *Le Climat de l'Algérie*, Institut de météorologie et de physique du globe de l'Algérie, Université d'Alger, Alger, 1946.

6. Le Houérou, H. N., Interannual variability of rainfall and its ecological and managerial consequences on natural vegetation, crops and livestock, in *Proc. 5th Int. Conf. on Mediterranean Ecosystems*, Paris, 1988, 323.

7. Le Houérou, H. N., Claudin, J., and et Pouget, M., Etude bioclimatique des steppes algériennes, *Bull. Soc. Hist. Nat. Afrique du Nord*, Alger, 68, 3, 1979.

8. Le Houérou, H. N., Classification écoclimatique des zones arides (s.l.) de l'Afrique du Nord, *Ecologia Mediterranea*, XV, 3, 1989.

9. Le Houérou, H. N., An outline of the bioclimatology of Libya, *Bull. Soc. Bot. Fr.*, 131, *Actual. Bot.* 2/3/4, 157, 1984.
10. Le Houérou, H. N. and Popov, G. F., An ecoclimatic classification of intertropical Africa, *FAO Plant Production and Protection Paper 31*, FAO, Rome, 1981.
11. Gounot, M. and Le Houérou, H. N., Carte bioclimatique de la Tunisie, 1 feuille 1/2 000 000eme (reeditions couleurs 1/500 000, Ann. INAT, 41, 1, 1959.
12. Le Houérou, H. N., La vegetation de la Tunisie steppique (avec references aux vegetations analogues d'Algerie de Libye et du Maroc), Ann. INAT, 42, 5: 1-624, 1 carte couleurs 1/500 000eme, 1969.
13. Gounot, M. and Le Houérou, H. N., Climatologie et bioclimatologie de la Tunisie septentrionale, Cartes et tableaux, *Ann. INRAT*, 41, 1, 1968.
14. Liquido, N. J., Shinoda, L. A., and Cunningham, R. T., Host plants of the Mediterranean fruit fly (Diptera: Tephritidae), an annotated world review, Misc. Publications, *Ent. Soc. America*, 77, 1991.
15. FAO, Date production and protection with special reference to North Africa and the Near East, *FAO Plant Production and Protection Paper 35*, FAO, Rome, 1982.
16. Lepesme, P., Les insectes des palmiers, Paris, 1947.
17. Back, E. A. and Pemberton, C. E., The Mediterranean fruit fly in Hawaii, *USDA Bulletin 536*, 1918.
18. Bodenheimer, F. S., Die Schädlingsfauna Palästinas, Mon. *Ang. Ent.*, Berlin, 10, 1930.
19. Bodenheimer, F. S., Citrus entomology in the Middle East with special references to Egypt, Iran, Irak, Palestine, Syria, Turkey, Junk, The Hague, 1951.
20. Avidov, Z. and Harpaz, I., *Plant Pests of Israel*, Israel Universities Press, Jerusalem, 1969.
21. Langronier, H., *Ceratitis capitata* dans les oasis sahariens, *Bull. Soc. Hist. Nat. Afrique du Nord*, Alger, 31, 7, 1941.
22. Blak, M., Neshnush, I. M., and Bin Kafu, A. A., New hosts of the fruit fly in Libya, *Agricultural Research Center Report*, Tripoli, 1987.
23. Sacantanis, K. B., La forêt d'arganier, le plus grand foyer de *Ceratitis capitata* Wied. connu au monde, *Service de la défense des végétaux*, Marrakech, Maroc, 1955.
24. Hendrichs, J. and Hendrichs, M. A., Mediterranean fruit fly (Diptera: Tephritidae) in nature: Location and diel pattern of feeding and other activities on fruiting and non-fruiting hosts and non-hosts, *Ann. Entomol. Soc. Am.*, 83, 632, 1990.
25. Bouhlier, R., De Francolini, J., and Perret, J., Essais attractifs pour la destruction de *Ceratitis capitata* Wied., *Revue Zoologie Agricole et Appliquée* 34, 149, 1935.
26. Long, G. H., Contribution à l'étude de la végétation de la Tunisie centrale, *Ann. Serv. Bot. & Agron. de Tunisie*, 27, 388, 1954.
27. Delanoue, P. and Soria, F., Les fruits de l'oranger amer (*Citrus bigaradia* Risso) réserve dangereuse en Tunisie de mouches des fruits (*Ceratitis capitata* Wied.), *Ann. INRAT*, 35, 185, 1962.
28. Di Cairano Vitale, M., La lotta contro la mosca della frutta, *Boll. Regio Ufficio Servizi Agrari*, Tripoli, 2, 7-8, 3, 1933.
29. Monjauze, A. and Le Houérou, H. N., Le rôle des Opuntia dans l'économie agricole nord africaine, *Bull. Ecole Nat. Supérieure d'Agriculture de Tunis*, 8-9, 85, 1965
30. Soria, F., Plantes hôtes secondaires de *Ceratitis capitata* Wied. en Tunisie, *Ann. INRAT*, 35, 51, 1962.
31. Le Houérou, H. N., unpublished data.

32. **Bounfour, M., El-Ayan, S., and Merhaben, J.,** Second meeting of National Coordinators, Casablanca, Morocco, 8-9 November 1990, Project RAF/5/013: Survey on the extent of medfly infestation in North Africa, *IAEA Report No. NC/2/90*, Vienna, 1990.

33. **Rieuf, P.,** Les champignons de l'arganier, Les cahiers de la recherche agronomique, 15, Direction de la recherche agronomique et de l'enseignement agricole, Rabat, Maroc, 1962.

34. **Voegelé, J.,** La cératite dans la région de Marrakech, Thèse présentée à l'Ecole Marocaine d'Agriculture, Meknès, 1953.

35. **Mazih, A.,** personal communication.

36. **Coupin, A.,** A propos de fruits véreux, Feuilles Inf. Vitic. Arbor, Tunisie, Tunis, 21, 3, 1950.

37. **Delanoue, P.,** Les nouvelles méthodes de lutte contre la mouche des fruits, *Ceratitis capitata* Wied., *IIIe Congrès Internat. Agrumiculture*, 1954.

38. **Bodenheimer, F. S.,** *Ceratitis capitata* in the Mediterranean Basin, EPPO, 1963, in *Report of the International Conference on the Mediterranean Fruit Fly and San José Scale*, Grünberg, A., Ed., Paris, 1963.

39. **Klein, H. Z. and Parker, M.,** Biological Studies on the Mediterranean Fruit Fly in the Jordan Valley, *Bull. Agric. Res. Sta.*, Rehovot, 32, 1942.

40. **Harris, E. J.,** Mediterranean fruit fly ecology in Morocco: Past and present, submitted, 1991.

41. **Meats, A.,** Abiotic mortality factors-temperature, in *Fruit Flies, Their Biology, Natural Enemies and Control, World Crop Pests*, Vol. 3B, Robinson, A. S. and Hooper, G., Eds., Elsevier, Amsterdam, 1989, 229.

42. **Delrio, G., Conti, B., and Crovetti, A.,** Effect of abiotic factors on *Ceratitis capitata* (Wied.) (Diptera: Tephritidae), I. Egg development under constant temperatures, in *Fruit Flies of Economic Importance* 84, Commission Eur. Communities, IOBC, A. Balkema, Rotterdam, 1984, 133.

43. **Crovetti, A., Conti, B., and Delrio, G.,** Effect of abiotic factors on *Ceratitis capitata* (Wied.) (Diptera: Tephritidae) II, Pupal development under constant temperatures, in *Fruit Flies of Economic Importance* 84, Commission Eur. Communities, IOBC, A. Balkema, Rotterdam, 1984, 141.

44. **EPPO,** Report of the International Conference on the Mediterranean Fruit Fly and San José Scale, Paris, 1963.

45. **Wong, T. T. Y., Whitehead, L. C., Kobayashi, R. M., Ohinata, K., Tanaka, N., and Harris, E. J.,** Mediterranean fruit fly: Dispersal of wild and irradiated and untreated laboratory-reared males, *Environmental Entomology*, 11, 339, 1982.

46. **Soria, F. and Cline, J. F.,** Etude du vagabondage de *Ceratitis capitata* Wied. en Tunisie à l'aide de radioisotopes, *Ann. INRAT*, 32, 79, 1962.

47. **Soria, F.,** Etude des populations et de dispersion de *Ceratitis capitata* Wied. (Dipt. Trypetidae) en Tunisie à l'aide des radioisotopes, in *Proceedings Symposium on Radiation and Radioisotopes Applied to Insects of Agricultural Importance*, IAEA, 1963, 357.

48. **Hafez, M., Abdel Malek, A. A., Wakid, A. M., and Shoukry, A.,** Studies on some ecological factors affecting the control of the Mediterranean fruit fly, *Ceratitis capitata* Wied. in Egypt by the use of the sterile male technique, *Zeitschrift Ang. Entomologie*, 73, 230, 1973.

49. **Wakid, A. M. and Shoukry, A.,** Dispersal and flight range of the Mediterranean fruit fly, *Ceratitis capitata* Wied. in Egypt, *Zeitschrift Ang. Entomologie*, 81, 214, 1976.

50. **Fletcher, B. S.,** Life history strategies of Tephritid fruit flies, in *Fruit flies, their Biology, Natural Enemies and Control, World Crop Pests*, Vol. 3B, Robinson, A. S. and Hooper, G. Eds., Elsevier, Amsterdam, 1989a, 195.

51. **Fletcher, B. S.,** Movements of Tephritid fruit flies, in *Fruit flies, their Biology, Natural Enemies and Control*, World Crop Pests, Vol. 3B, Robinson, A. S. and Hooper, G. Eds., Elsevier, Amsterdam, 1989b, 209.

52. **Steiner, L. F., Mitchell, W. C., and Baumhover, A. H.,** Progress of fruit fly control by irradiation sterilization in Hawaii and Mariana Islands, *Intern. Journal of Applied Radiation and Isotopes*, 13, 427, 1962.

53. **de Murtas, I. D., Cirio, U., and Enkerlin, D.,** Dispersion of *Ceratitis capitata* Wied. on Procida Island, I. Distribution of sterilized Mediterranean fruit fly on Procida and its relation to radiation doses and feeding, *European Plant Protection Bull.*, 6, 63, 1972a.

54. **de Murtas, I. D., Cirio, U., and Enkerlin, D.,** Dispersion de *Ceratitis capitata* Wied. dans l'île de Procida, II. Etude des déplacements, *European Plant Protection Bull.*, 6, 69, 1972b.

55. **Harris, E. J. and Olalquiaga, G.,** Occurrence and distribution patterns of Mediterranean fruit fly (Diptera: Tephritidae) in desert areas in Chile and Peru, *Environ. Entomol.* 20, 174. 1991.

56. **FAO,** Africa, Statistical Basebook for Food and Agriculture, FAO, Rome, 1986.

57. **Buyckx, E. J. and Vita, G.,** Report of visits to Algeria, Morocco and Tunisia, Project RAF/5/013: Survey on the extent of medfly infestation in North Africa, IAEA, Vienna, 1988.

58. **El-Sayed, A. M. K., Darwish, Y. A., and Mannaa, S. A.,** Evaluation of certain trapping systems for capturing Mediterranean fruit fly, *Ceratitis capitata* (Wiedemann) during seasons of 1987 to 1991 in Assiut, Egypt, Mimeographed Report, Plant Protection Dept., Faculty of Agriculture, Assiut University, Assiut, 1991.

Chapter 13

FRUIT FLY FREE AREAS: STRATEGIES TO DEVELOP THEM

Aldo Malavasi, G. Greg Rohwer, and D. Scot Campbell

I. INTRODUCTION

A major impediment to fruit and vegetable crop diversification in most countries is the presence of several fruit fly species of economic and quarantine importance. These fruit flies are a concern to many noninfested importing countries. Potential markets are available for numerous host fruits and vegetables that are not allowed entry because of importing country fruit fly quarantine restrictions.

Species of economically important fruit flies are of great concern to most countries and are particularly important to developing countries in the production, in-country marketing, and export of host fruits and vegetables. This is true for fruit fly species that pose a threat to suitable habitats in the entire world. The principal means of dealing with foreign species is through quarantine actions imposed by noninfested countries and by conducting early detection and eradication programs.[1] Many countries take actions to eradicate incipient introductions when possible or to establish management programs to reduce populations and prevent spread.[2]

Fruit fly spread was minimal prior to the development of modern transportation systems. However, perishable host materials are now rapidly transported from infested to noninfested areas. Also, travelers often carry infested host materials in their carry-on and checked luggage, which complicates enforcement of quarantine restrictions.

Chemicals such as ethylene dibromide and methyl bromide have been used for decades as quarantine treatment against all fruit fly species. In operational terms, fumigation was a very convenient procedure for disinfecting hosts for export. However, EDB and MB can no longer be used. Tests with rats at maximum tolerable doses caused cancer in the surviving test animals. Under United States laws such chemicals are cancelled even though there is no dose response relationship. The worldwide growing concern in the use of pesticides in general and particularly in food, associated with such toxicological studies, resulted in banning the use of chemicals normally employed in fresh fruit disinfestation. In large degree, the certification of commodities through fly free areas and other alternative systems have been developed as a means of facilitating international commerce in certain produce that might otherwise be banned.

Some of the alternative treatments that have been approved for fruit fly disinfestation require intensive research in the country of origin for approval of complex or sophisticated treatments.   The research must be conducted in the country of origin because the host fruit fly species system is generally characteristic of the country.   After the approval of the quarantine treatment, costly facilities must be built to meet the requirements for heat, cold, and irradiation treatments.   Many developing countries can neither afford to undertake the research nor can they provide the types of facilities required.

One of the most promising approaches is the concept of fruit fly free areas, which allows for export of crops produced in the area to importing countries without post-harvest treatment.   The concept of fly-free areas involves the use of limited resources in the development of good pest management practices and improvement of production and quality of fruit.

II.   FRUIT FLY FREE AREA CONCEPT

A fruit fly-free area is an area with no detectable populations of a particular fruit fly species.   It is usually separated from infested areas by natural or artificial barriers and protected by quarantine as necessary.

There are two models of free areas shown in Figure 1: (1) the *fly-free zone*, also called designated area, where an entire geographic or political entity is recognized to be free from the target fruit fly species; and (2) the *fly-free production field(s)* that uses the concept of systems approach involving a series of actions to manage a pest so that it is absent from specific parcels or production areas but does not necessarily involve freedom from large geographic or political entities.

In the first model, eradication treatments are applied as necessary throughout the area and regulatory protection is provided to the entire area. Two good examples of fly free zones are in the State of Sonora in northwest Mexico and for most areas in Chile.   In Sonora, eradication treatments resulted in freedom of all fruit fly species of quarantine importance.   Chile is considered free of the Mediterranean fruit fly (medfly), *Ceratitis capitata* (Wied.), and other economic species of fruit fly of the hosts cleared for export.   Other examples are in Mossoro in the northeast part of Brazil and Guayaquil in the southeast part of Ecuador.   Both are considered free of *Anastrepha grandis*, the South American cucurbit fly and other economic species that attack melons.

In the second model, fly free production field(s) have been developed in recent years and although the presence of fruit flies might be detected a few kilometers away, specific production areas are accepted as free. According to this concept, a fly-free field can be established in small areas such as a farm, an orchard, or a group of properties and involve the certification of specific production fields.[3]   Quarantine protection of these

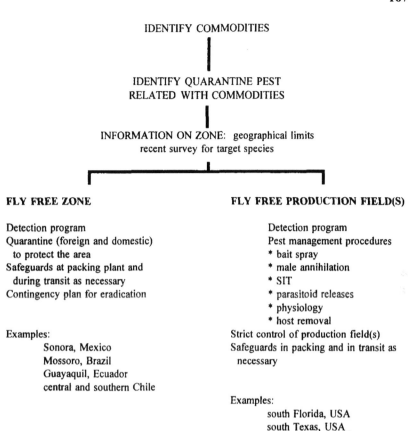

IDENTIFY COMMODITIES

IDENTIFY QUARANTINE PEST
RELATED WITH COMMODITIES

INFORMATION ON ZONE: geographical limits
recent survey for target species

FLY FREE ZONE

Detection program
Quarantine (foreign and domestic)
 to protect the area
Safeguards at packing plant and
 during transit as necessary
Contingency plan for eradication

Examples:
 Sonora, Mexico
 Mossoro, Brazil
 Guayaquil, Ecuador
 central and southern Chile

FLY FREE PRODUCTION FIELD(S)

Detection program
Pest management procedures
 * bait spray
 * male annihilation
 * SIT
 * parasitoid releases
 * physiology
 * host removal
Strict control of production field(s)
Safeguards in packing and in transit as
 necessary

Examples:
 south Florida, USA
 south Texas, USA
 San Joaquin Valley, CA, USA
 Hawaii, USA

Figure 1. Requirements regarding the two models of fruit fly areas.

zones is provided by the growers and is not necessarily required by the country. Examples of this model are in south Texas where the program certifies citrus production groves as being free of the Mexican fruit fly, *Anastrepha ludens* (Loew), and central Florida that recognizes specific citrus groves as free of the Caribbean fruit fly, *A. suspensa* (Loew).[4,5]

Fly-free areas of both models may be on the edge of the geographic distribution of the fruit fly species or in biologically marginal areas of the target species. Where they are in favorable biological areas, the species has been eradicated if introduced or never found in the area and quarantine protection is applied to prevent entry.

The concept of a pest free area is more advantageous if: (1) the importing country determined that the target species is an important regulatory pest; (2) post-harvest treatments for the commodity were not satisfactory or unavailable; or (3) the establishment of a fly free area was biologically, economically, and politically viable.

The definition of a free area is based on three major points: (1) area delineation; (2) target pest species or group of species; and (3) one or more host commodities produced in the area.

III. CHARACTERIZATION OF A FREE AREA

In the biological sense, there are two types of free areas, a "natural" and a "managed" free area. A natural fly-free area is that in which the species does not naturally occur because of ecology, host preference, geographical distribution, etc. In this case, the entire production zone is encompassed by the fly-free area. Managed fly-free areas, on the other hand, are production zones from which the target fruit fly has been eradicated. Primary hosts have been found infested previously and control and eradication measures were taken against the species. However, it is clear and it can be anticipated that the pest will establish easily in these areas because there are no limiting factors to prevent its establishment. In this case, the area must be protected by regulatory action and trapping must be conducted to detect any introductions.

An area could be considered fly-free for an ecological or biological reason, such as lack of host material. When the area becomes an agriculture frontier with new crops being cultivated, there is a risk of bringing infested material into the area. Agricultural expansion usually brings rapid urbanization that, in a few years, results in patchy host distribution in gardens, backyards, and domestic orchards. The presence of ornamental and fruit trees that are fruit fly hosts in urban areas typifies a major constraint in detection and eradication programs. In the Los Angeles basin, host material presence in gardens and backyards represents the major single factor that complicates the eradication programs. One requirement for the Caribbean fruit fly free zone in Florida is that the production area is located at least 4.8 kilometers from urban areas[3] where there are favorable hosts of the pest species.

Secondary hosts may serve as a natural repository for fruit flies, maintaining the population at very low densities. A few gravid females could infest a large number of favored fruits causing the population to increase dramatically in the next generation.[6]

If the fly-free area is not ecologically favorable for the establishment of a given species, a large planting of hosts creates a good condition for colonization. In this case, the traffic of infested material into the area is the most important cause of introduction. Inspection at road stations on major highways, airports and ports is necessary to prevent the introduction of infested hosts.

The establishment of fly free areas with high human populations or areas crossed by many highways, roads, and waterways will be more difficult because of the potential for introduction. The costs of full inspections at roads and airports might not be feasible in many areas in developing countries. In such situations, only the systems approach may be feasible.

The following aspects are considered to characterize and establish an area as a fly free area:

A. **PHYSICAL**
   . geographical barrier
   . adverse climatic conditions
B. **ECOLOGICAL**
   . target pest species
   . primary host presence
   . secondary host abundance
C. **BIOLOGICAL (PEST)**
   . host ranking
   . life cycle
   . bionomics
   . behavior
D. **POLITICAL, ECONOMIC, AND SOCIAL**
   . crop production value
   . economic feasibility
   . national and local government support
   . regulation of movement of host material
   . requirements of importing countries or states
   . public support

A. **PHYSICAL**
   A fly free area must be isolated from other areas and/or adequately protected by quarantine as necessary. The most common natural barrier is the ocean. Many oceanic islands have been established as free of flies in Japan,[7] the Caribbean,[8] and in Chile.[9] Other potentially fly-free islands are located in Hawaii,[10] Southeast Asia, Malaysia, the Philippines, Thailand, and Indonesia. It was not by chance that the first eradication program of the screwworm was carried out in the Curaçao Island in the early 50's.[11] Islands provide enough isolation to avoid migration by natural dispersion, although spread may occur during severe weather conditions. Strong quarantine measures at the ports of entry, by sea and by air are necessary to prevent artificial introduction and reestablishment.

Deserts and mountains provide natural barriers for long distance flights of the pests. The best example of deserts and mountains being barriers for fruit fly migration is found in Chile. Medfly and *Anastrepha* species are found in northern Chile, in the Tacna Valley east of the Andean Cordillera. The Arrica desert and the Cordillera provide a strong obstacle to medfly migration. Thus, central and southern Chile are free of medfly. Deserts are also the physical basis for the establishment of the Sonora fly-free zone in northwestern Mexico and the *Anastrepha grandis* fly-free zone in northeast Brazil.

Adverse temperature range is probably the most important single factor affecting the distribution of fruit flies in the world. Fruit flies indigenous to tropical and subtropical regions, including the genera *Ceratitis*, *Anastrepha*, and *Bactrocera*, will not overwinter in cold climates. Northern European countries do not have quarantine restrictions regarding tephritids when importing fresh fruits produced in infested countries, basically because most countries in that part of Europe have cold winters.

Low humidity in dry areas also is a factor influencing the distribution pattern of fruit flies and should be considered in the establishment of a fly-free area.

B.   ECOLOGICAL

It is necessary to define which species or group of species will be the target for the establishment of a fly free area. The knowledge of geographical and spatial distribution of tephritids is very inconsistent for vast areas in tropical and subtropical regions. If the occurrence of the target species in the projected fly-free area is not well determined, the first step should be an intensive survey to verify its distribution.

At the same time, a survey to quantify the presence of primary hosts must be carried out to estimate the availability of oviposition sites. The sequence of different hosts is critical for tropical species to maintain the population density at a minimal level. When there is scarcity of primary hosts, these species shift to secondary hosts, which then operate as a natural repository in sustaining the population.[12]

C.   BIOLOGICAL

Lack of information on the biology of the target species can jeopardize any monitoring and eradication program. Biological data on host range and host ranking are crucial to plan the strategies for establishing fly-free areas. Intensive fruit collecting in the proposed zone may be essential to determine what fruit species should be cataloged as hosts. Data from other countries, or even the same country but in a diverse ecological situation, should not be accepted because of the tremendous geographical variation on host range and rank in tropical tephritids. For instance, carambola is not a host for medfly in southern Brazil but is highly infested in the northeast, only 2,000 kilometers apart.[13]

Bionomic and demographic data are required to explain the population fluctuation of immature stages and adults and then in planning the best tools and schedule for monitoring.[14] Information on fly behavior is important for trap location, for intra- and inter-cropping and for determining suppression and eradication techniques to be used.

D. POLITICAL, ECONOMIC, AND SOCIAL

The first point when a free area is being considered involves economic feasibility. If the technology for monitoring, survey, and eradication is expensive, a cost/benefit analysis should be conducted. The potential value of the crop, including expanded exports, should be enough to justify the establishment and maintenance of the fly free area.

Strong interaction within the plant health community is essential in the establishment and maintenance of fly-free areas. In this sense, the plant health community is the industry (producer and shipper), regulatory, research, and extension entity. All of them should be directly involved and committed to the fly-free area concept. Each entity has an important role that cannot be delegated without risking program success.

Support by local and national governments is essential to enforce the regulations that control the fruit traffic through the roads in the zone, to provide technical personnel for the program, and to be the interlocutor with plant protection agencies of the importing countries.

Because the ultimate objectives are to establish and maintain a pest-free area, such programs, depending upon their magnitude and location, may affect the general public. In this sense, social considerations play an important role in the successful implementation and maintenance of a fly-free area. Persons living in the area must cooperate with the program by not bringing infested material into the area, by not planting or growing host plants as outlined in the overall plan and by accepting the measures implemented such as bait spraying, and host tree removal where or when necessary. Neglecting an explanation about the meaning of the program to the general public with its social and economic benefits could jeopardize it. Public awareness campaigns have been used in the Mexico and Guatemala medfly eradication programs with success. It is the responsibility of the extension community for education and public relations. Growers and grower organizations can be of great assistance especially in regard to urging compliance with quarantine and program requirements.

IV. STRATEGY

The steps to develop and maintain a fruit fly-free area are summarized in Table 1.

Table 1

**Major Steps for the Establishment and Maintenance
of a Fruit Fly Free Area**

	SURVEY	SURVEY TRAPPING		LARVAL	
		Establishment	Maintenance	Establishment	Maintenance
Free zone (designated areas)	Entire zone	Negative Eradicate if finds in zone	Eradicate if finds in zone	May be required dependent on trap efficacy and knowledge host susceptibility	May be desirable to spot check highly preferred hosts
Free Production Field(s)	Entire fields and environs and around packing sheds	Negative If positive, eradicate fields and environs in and around packing sheds	If finds, eradicate fields and environs and around packing sheds	Same as above for free zone	Same as above for free zone

MANAGEMENT

	Establish	Maintain
Free Zone	Eradicate any infestation in entire zone	Eradicate any incipient infestation in entire zone
Free Production Field(s)	Eradicate any infestation in field(s) and environs and around packing sheds Size environs dependent on trap efficacy and host susceptibility	Eradicate any infestation in field(s) and environs and around packing sheds

QUARANTINE

	Formal Quarantine	Road Stations	Packing	Packing Plant to Port of Export
Free Zone	Document of free zone involving federal government	Usually required	In free zone and safeguards to prevent hitchhikers	Protect (screen) to prevent fly entry between packing plant and port of export
Free Production Field(s)	Not required (grower control)	Not required	Host removal in environs or treatment and protection from hitchhikers	Same as for free zone

A.   IDENTIFYING COMMODITIES

A fly-free area is usually established when there is a large production of a commodity that growers want to export or there is a potential for new markets when a free area is established.   In this sense, first the land is occupied by the crop, and subsequently, the fly-free area concept is applied. Unfortunately, the expansion of agriculture usually takes into account the edaphic, climatic, and market conditions, but not the presence of pests.

In the area that potentially might be considered fly-free, it is necessary to identify the specific commodities that are candidates for export.   If a commodity is a potential fruit fly host, its status should be determined for the tephritids occurring in the area.   Biological data is not available in the scientific literature in many cases, therefore pest/host relationships must be established.

The importing countries are also concerned with other quarantine insects that may be sent in a fruit shipment.   Exclusion of miscellaneous (hitchhiking) pests of potential quarantine importance must be provided at least for the fruit shipment itself.   This may require modification of packing facilities, transportation systems, or an inspection prior to export.

B.   IDENTIFYING QUARANTINE PESTS RELATED WITH COMMODITIES

There is a list of frugivorous insects that are associated with each fruit species in terms of infestation.   There is, however, a lack of information on host range for many species of fruit flies.   The scientific data that provide material for the plant protection agencies to make decisions are scarce and only partially meet their needs.   When the information about a commodity and a given fruit fly species is not consistent or is incomplete, plant protection agencies usually take a more drastic approach by considering a secondary species as a quarantine pest.

The host range list of fruit flies is the most important tool for making decisions on quarantine pests.   Many countries have well organized lists of quarantine pests related with fruits.   Contact among the national plant protection agencies and the harmonization of policies have positively affected the regulations, making them more congruent even in different countries and regions.

C.   INFORMATION ON FLY-FREE AREAS
1.   Geographic Limits

The boundaries of fly-free areas must include the entire host production in the zone and be well defined.   The limits may be rivers, lakes, mountains, deserts, or artificial features such as highways.

2.   Recent Survey on Target Species

Surveys are necessary to provide importing countries with information about the presence or absence of target species in the area. Adequate survey of adults and immature stages is a major requirement before the importing country recognizes the establishment of a fly-free area. If the surveys show that the production area remains free from infestation by the target species, the area may be eligible as a fly-free area.

Detecting immature stages in fruit is more costly and time-consuming than detecting adults with traps, but this method should be considered in certain situations including cutting culled fruits at packing sheds. Presently, only fruit cutting has been employed to detect infested fruits. Thousands of fruits must be cut during some host seasons to confirm the absence of infestations. This work is very tedious, and because the expectation is for a negative result, this likely will be fulfilled. A promising new technology involves acoustic detection of larvae. Fruit fly larvae emit a characteristic sound while feeding that can be detected with highly sensitive microphones.[15] The potential efficiency of this method far surpasses fruit cutting.

Holding fruit to obtain emerging larvae is another method to detect a fruit fly population. This method is also costly, demands a sampling system to collect fruits, and space for holding them. It has been used only in a limited way to detect populations and in conjunction with fruit cutting. Both methodologies are important precursors to the development of fly-free areas, especially in situations where host susceptibility is not known.

In areas where there is no knowledge of a fruit fly population, an intensive survey is necessary for a sufficient time period covering an agreed upon number of life cycles of a pest — usually three or more. The time period will be influenced by the efficacy of the detection system.

D.   MAINTENANCE OF THE FLY-FREE AREA

The agencies in charge of a fly-free area should have a four-part program: exclusion, detection, eradication/management, and public information. The exclusion program prevents the introduction of the target species through the use of inspection stations if the pest is present in the country. An adequate program at ports of entry should be included in situations where the pest is absent from the country or large portions thereof such as in Chile. The detection program is to find as early as possible, any incipient infestation in the area, while the eradication program is to eliminate the outbreak and/or manage the population at acceptable levels. The public awareness program is to inform people of the importance of the program and obtain their cooperation by not bringing host material into the area from areas where the pest occurs. The California Department of Food and Agriculture uses this system to protect the state against the introduction of exotic pests.[16]

There are two possible situations for dealing with populations of fruit flies in terms of a fly-free area: the fly-free zone or the fly-free production field. When the target species is absent from the zone, all available measures are used to exclude the flies. In the second case, a population of the target species occurs nearby the production field and measures should be used to suppress and control it to prevent it from dispersing. The strategies are different in each case, although some measures are common to both systems.

1.   Fly Free Zone Model

In some cases, for physical, biological and/or ecological reasons, the population of the target species is not established. For example, in Chile *C. capitata* occurs only in the far northern part of the country, on the border with Peru. Production areas of fruit are isolated from the north by the Arrica desert. This provides an excellent barrier to medfly migration. In this situation, measures would be concentrated to prevent artificial movement of adults and immature stages in commerce or by homeowners through infested fruits.

a.   Detection Programs

A well planned fruit fly detection and quality control program is the fulcrum for any fly-free area maintenance. However, as it was well pointed out by Cunningham,[17] detection trapping is a thankless job. "It brings either no news or only bad news." The opposite viewpoint regarding detection is that it provides pertinent information on pest populations and through continued negative surveys enhances exports and affords the potential for increasing food production through better pest management systems.

Because of the peculiarity of detection trapping, quality control is quite important. In a good program, a significant component is the validation that survey guidelines have been properly implemented and are followed.

There are international efforts to standardize fruit fly detection. Entomologists have failed to provide data for answering questions regarding detection of fruit flies.[17] Over the years, different trap models have been exhaustively tested, but only two basic models are widely used and approved by the regulatory agencies: the Jackson trap for the medfly, Oriental fruit fly, *B. dorsalis* (Hendel), melon fly, *B. cucurbitae* (Coquillett), and the McPhail trap for *Anastrepha* spp. The attractants used in these traps are: trimedlure for medfly and natal fruit fly, *Ceratitis rosa*, methyl eugenol for Oriental fruit fly, cue-lure for melon and Queensland fruit flies, and food lures for *Anastrepha* spp. Torula yeast and protein hydrolysate are the most common attractants used in McPhail traps. However, if a fruit fly species is considered as a quarantine pest, additional tests must be conducted to determine trap models and attractants. For instance, until recently there was a total lack of information for trapping *A. grandis*, a species that attacks cucurbits in South America. During the process for the establishment of an *A. grandis* fly-free zone in Brazil, research was conducted to determine what attractant could be used as bait and the trap model.[18,19]

The trapping densities utilized to establish and maintain the integrity of the free areas should be based on research findings and are established by mutually acceptable work plans developed by appropriate plant protection officials from the importing and exporting countries. Trap array and servicing are crucial to the successful establishment of a fly-free zone, and subsequently, to maintaining the zone's status. The Animal and Plant Health Inspection Service of the United States Department of Agriculture (APHIS/USDA) has established recommended trapping densities and lures for a variety of situations.[20] While these and other trapping plans are valuable starting points in any work with fruit flies, any trapping guidelines should be under continuous review. Further, they should be adjusted to reflect current technology and actual working situations in the area being considered for fly free status. In any event, the traps should be installed in the most preferred fruiting host trees in urban areas or in commercial host situations to maximize the probability of detection.

If the target fruit fly species is detected, immediate notification is issued to the importing country. All shipments are interrupted and actions outlined in the contingency plan are immediately initiated. Such actions are outlined in the work plan for acceptance and maintenance of the fly-free area.

b.   *Traffic Control and Incoming Transit*

Prevention of introductions of target species into a non-infested area is usually more economical and feasible than eradication after fruit flies have become established. In this case, inspection sites at ports and airports and at road stations are crucial in preventing the introduction of infested material into the area.

Most fly-free zones are in agricultural areas where intensive transit of commodities and passengers occur. In many countries, bringing fruits and vegetables to relatives is part of the culture. This traffic represents the most important threat to a fly-free area. The commerce between fly-free and infested zones must be controlled through installation of road stations along major highways. Public awareness of the program is necessary to obtain cooperation to prevent introduction of exotic species and to facilitate inspections at all points.

c.   *Contingency Plan of Eradication*

If the exclusion program fails to prevent the introduction of target species and the detection system discovers an infestation, a contingency plan for eradication must be ready. This plan should use the best tools available (bait spray, male annihilation, etc. - see discussion below) to knock down and eliminate the established population.

d. Examples

Fruits produced in central and south Chile can be exported to the United States using the fly free zone concept for medfly and *Anastrepha* spp. Chile conducts a detection program for all economic species of fruit flies. The state of Sonora in northern Mexico is considered a fly-free zone for *Anastrepha* spp. and other economically important species. Citrus, stone fruits, and apples can be exported from this area without post-harvest treatment. The Mossoro area in the northern region of Brazil and Guayaquil in Ecuador are in the process of being recognized by APHIS/USDA as free of *Anastrepha grandis* that infest cucurbits.[21] Honeydew melons produced in both areas are eligible for export to the United States. In the absence of a free zone, entry would be prohibited because there are no available commodity treatments. These cases illustrate situations where the target species is absent from the area and necessary quarantine is taken to prevent artificial introductions. Surveys agreed upon are conducted to promptly defeat any introductions.

2. Fly Free Production Field(s)

In some areas, it is possible to design programs that allow the movements of hosts from production field(s) without post-harvest commodity treatment. Protocols agreed upon between exporting/importing countries provide for detection surveys to assure pest freedom in the field(s). A specified buffer zone around production field(s) is required in which hosts may not be present, or agreed upon suppression measures are applied. Procedures are outlined which allow for packing of the hosts in a manner to prevent spread. This is referred to as a systems approach because the geographical area is not free of the pest but the agreed upon procedures prevent host infestation in the specified field(s). It is particularly applicable in situations where the export crop is not a preferred host. Formal quarantines to protect the production field(s) may not be necessary. Growers take necessary actions to prevent introductions into the field(s).

In the systems approach, the program uses all scientific and historical data available regarding the species and the commodity. For instance, for *A. ludens* in Texas and *A. suspensa* in Florida, information was reviewed reporting the host status of citrus, data on capture of flies, distribution of wild hosts, etc.[3,5]

a. Suppression and Control Measures

Pest suppression technology should be used to assure that the fruit fly population is below the risk level. The most important tools used for suppression are the use of bait sprays, release of sterile males, and male annihilation. These traditional approaches have worked for decades in areas such as south Texas and Florida.[3] New tools are being developed and will be useful for such programs. These new technologies include inundative releases of parasitoids to suppress adult population of tephritids. This method has been used in Hawaii[22] against three species, medfly, oriental, and melon fly; in Florida for suppression of Caribbean fruit fly, *A. suspensa*;[23] and for control of the *Anastrepha* complex occurring in Mexico. Biocontrol through pathogens such as *Bacillus thuringiensis* and nematodes as *Steinernema carpocapsae*,[24] have been used against medflies in Hawaii. The use of gibberellic acid for prophylaxis against infestation of citrus fruit by *Anastrepha* spp. and medfly[25] are currently under investigation in Florida, Brazil, and Mexico. The combination of different technologies such as augmentative parasitoid releases plus SIT or male annihilation is a promising approach.

b.   *Examples*

Citrus produced in designated Caribbean fruit fly-free fields in Florida can be shipped to other states in the United States and to Japan if specific protocols are followed.[3,5] Specific production fields in the Rio Grande Valley of Texas are considered free of the Mexican fruit fly, *A. ludens*, and citrus is certified if produced in specific areas where the fly is absent. The San Joaquin Valley in California is considered free of the walnut husk fly, *Rhagoletis completa*, before July 1 each year, because the species is univoltine and its emergence is predicted to occur after July 1.[26]

V. CONCLUSION

Current programs of fruit fly-free areas have been adequately developed and have benefited both exporters and importers. The operation of fly-free areas utilizing the same techniques that have been developed for eradication programs has facilitated their establishment. New tools available in coming years should provide faster development of the fly-free area concept.

It would benefit the agricultural economy in each country to evaluate the fruit fly situations and to design fly-free area programs that would improve crop production and export. Preliminary information regarding pest distribution and geographical isolation is available, indicating that the fly free area concept or systems approach to management could be developed to include many areas of the world.

Although this chapter has been targeted toward fruit flies, it is also possible for the concept to be applied to other pests and diseases such as citrus canker, khapra beetle, etc.

The use of the concept of a fly-free area should have implications on commerce among countries. Current discussions are occurring under GATT regarding the use of the pest-free area concept. Furthermore, the regional plant protection organizations and FAO are working on various aspects to standardize procedures at the global level. The utilization of the fly-free area should be one of the procedures that is recognized internationally as a certification procedure which provides adequate protection to importing countries.

REFERENCES

1. Rohwer, G. G., Recommendations regarding fruit fly management/eradication in the Western Hemisphere, OIRSA/NAPPO, 1992.

2. Klassen, W., *Eradication of introduced arthropod pests: theory and historical practice*, Entomol. Soc. of America, Lanham, Maryland, 1989, 29 pp.

3. Reiherd, C., Nguyen, R., and Brazzel, J. R., Pest-free areas, in *Quarantine Regulations*, Sharp, J. L., Ed. (In Press).

4. Reiherd, C., Citrus production areas maintained free of Caribbean fruit fly for export certification, in *Proc. Intl. Symp. on Fruit Flies of Economic Importance, Antigua Guatemala, 1990*, Liedo, P. and Aluja, M., Eds., Springer-Verlag, 1992, 407.

5. Simpson, S. E., Development of the Caribbean fruit fly-free zone certification protocol in Florida, *Fla. Entomol.*, 76, 228, 1993.

6. Harris, E. J., Hawaiian islands and North America, in *Fruit Flies, Their Biology, Natural Enemies and Control*, Vol. 3A, Robinson, A. S. and Hooper, G., Eds., Elsevier, Amsterdam, 1989, chap. 2.6.

7. Kawasaki, K., Eradication of fruit flies, in *Proc. Int. Symp. Biol. Control Fruit Flies*, Ginowan, Okinawa, Japan, 1991.

8. Hilburn, D. J. and Dow, R., Mediterranean fruit fly (*Ceratitis capitata*) eradication from Bermuda, *Fla. Entomol.*, 73, 342, 1990.

9. Bateman, M. A., Insunza, V., and Arretz, P., The eradication of queensland fruit fly from Easter island. *FAO Plant Protection Bulletin*, 21, 114, 1973.

10. Vargas, R. I. and Spencer, J. P., Hawaii fruit fly eradication pilot tests, *Proc. Int. Symp. Biol. Control Fruit Flies*, Ginowan, Okinawa, Japan, 1991.

11. Knipling, E. F., Eradication of plant pests-Pro. Advances in technology for insect population eradication and suppression, *Bul. Entomol. Soc. Am.*, 24, 44, 1978.

12. Malavasi, A. and Morgante, J. S., Adult and larval population fluctuation of *Anastrepha fraterculus* and its relationships to host availability, *Environ. Entomol.*, 10, 275, 1981.

13. Malavasi, A., unpublished data.

14. Carey, J. R., Demographic analysis of fruit flies, in *Fruit Flies, Their Biology, Natural Enemies and Control*, Vol. 3B, Robinson, A. S. and Hooper, G., Eds., Elsevier, Amsterdam, 1989, chap. 8.6.

15. Webb, J. C., Slaughter, D. C., and Litzkow, C. A., Acoustical system to detect larvae in infested commodities, *Fla. Entomol.*, 71, 492, 1988.

16. Dowell, R. V. and Krass, C. J., Exotic pests pose growing problem for California, *California Agriculture*, 46, 6, 1992.

17. **Cunningham, R. T.,** Population detection, in *Fruit Flies, Their Biology, Natural Enemies and Control,* Vol. 3B, Robinson, A. S. and Hooper, G., Eds., Elsevier, Amsterdam, 1989, chap. 7.1.

18. **Malavasi, A., Duarte, A. L., Cabrini, G., and Engelstein, M.,** Field evaluation of three baits for South American cucurbit fruit fly (Diptera:Tephritidae) using McPhail traps, *Fla. Entomol.,* 73, 510, 1990.

19. **Duarte, M. D., Amaral, P. M., and Malavasi, A.,** Comparison of glass and plastic McPhail traps in the capture of the South American fruit fly, *Anastrepha fraterculus* (Diptera:Tephritidae) in Brazil, *Fla. Entomol.,* 74, 467, 1991.

20. **Anonymous,** Surveys by foreign governments for exotic pests, Animal and Plant Health Inspection Service, United States Department of Agriculture, 1986.

21. **Nascimento, A. S., Malavasi, A., and Morgante, J. S.,** Programa de monitoramento de *Anastrepha grandis* (Macquart, 1845) (Diptera, Tephritidae) e aspectos de sua biologia, in *Moscas-das-frutas no Brasil,* Souza, H. M. L., Ed., Fundação Cargil, Campinas, Brazil, 1988, chap. 4.

22. **Wong, T. T. Y., Radaman, M. M., McInnis, D. O., Mochizuki, N., Nishimoto, J. I., and Herr, J. C.,** Augmentative releases of *Diachasmimorpha tryoni* (Hymenoptera: Braconidae) to suppress Mediterranean fruit fly (Diptera: Tephritidae) population in Kula, Maui, Hawaii, *J. Biological Control,* 1991.

23. **Sivinski, J.,** The influence of host fruit morphology on parasitization rates in the Caribbean fruit fly *Anastrepha suspensa, Entomophaga,* 36, 447, 1991.

24. **Lindegren, J. E., Wong. T. T., and McInnis, D. O.,** Response of Mediterranean fruit fly, melon fly and oriental fruit fly (Diptera: Tephritidae) to the entomophagous nematode, *Steinernema feltiae* in field tests in Hawaii, *Environ. Entomol.,* 19, 383, 1990.

25. **Greany, P. D., McDonald, R. E., Schroeder, W. J., and Shaw, P. E.,** Improvement in the efficacy of gibberellic acid treatments in reducing susceptibility of grapefruit to attack by the Caribbean fruit fly, *Anastrepha suspensa, Fla. Entomologist,* 74, 570, 1991.

26. **Yokoyama, V. Y., Miller, G. T., and Hartsell, P. L.,** Pest-free period and methyl bromide fumigation for control of walnut husk fly (Diptera: Tephritidae) in stone fruits exported to New Zealand, *J. Econ. Entomol.,* 85, 150, 1991.

Chapter 14

PINK BOLLWORM STERILE MOTH RELEASES: SUPPRESSION OF ESTABLISHED INFESTATIONS AND EXCLUSION FROM NONINFESTED AREAS

T. J. Henneberry

I. INTRODUCTION

A. BRIEF HISTORY

Sterile insect release methodology as conceived and implemented by Knipling[1] was considered a potential option for pink bollworm (PBW), *Pectinophora gossypiella* (Saunders), population suppression in the early 1960's.[2] The development of an artificial diet[3] and mass-rearing technology[4] were research contributions that supported the feasibility of the approach as a strategy to complement other methods in an integrated system to manage PBW populations. Subsequently, Graham[5] found that newly emerged PBW moths exposed to 25 krad of gamma radiation or more and crossed with untreated moths produced no fertile adult progeny. Sterile moth releases (15-40 krad) in field cages with untreated insects reduced developing PBW populations 72 to 91 percent over two generations.[6,7]

Irrespective of these promising results, the development of a workable sterile PBW release system was in its infancy when the first PBW's were found in the Palo Verde and the Bard Valleys of southern California in 1965. The magnitude of the PBW threat to the California cotton industry was quickly realized and prompted immediate action by state and federal agencies, the scientific community, and cotton growers. Emergency funding by the California Department of Food and Agriculture (CDFA) to insecticide treat 2024 cotton ha in the Bard Valley, CA, in late November, 1965, interim PBW quarantine and regulatory actions, establishment of a PBW Eradication Area Regulation, and contact with the U.S. Department of Agriculture to develop plans for PBW control, suppression, or eradication in California, Arizona, and Mexico were quickly accomplished.[8] The International PBW Work Committee was established in 1967 to consider long-range PBW management objectives on an areawide basis.[9] The emergency actions taken, in relation to preventing further spread of infestations, were fully appreciated when the first PBW moth was caught in the San Joaquin Valley on October 2, 1967 and a single larva recovered from an infested boll on November 3. The California Cotton Pest Control Board first met on May 25, 1967 following legislation appropriating state funds for PBW control and providing for a cotton bale fee (50 cents) to support the CDFA PBW eradication efforts. The U.S. Department of Agriculture's Entomology Research Division, Plant Pest Control Division, the National PBW Task Force, International PBW Work Committee, and the California Pest Control Board made a concerted effort to provide support for implementation of a pilot sterile PBW moth release program during the 1968 cotton-growing season.[8] The program was considered a temporary, stop-gap measure, to suppress population establishment in the San Joaquin Valley until effective cultural and biological control methods could be developed for southern desert cotton-growing areas.[10]

B.   BARRIER ZONE METHODOLOGY

The rationale for using PBW sterile moth releases to exclude or prevent infestation establishment was the result of experience gained in the southeastern screwworm, *Cochliomyia hominivorax* (Coquerel), eradication program.[11] One of the most important factors contributing to the success of that program and applicable to most large-area insect suppression programs involving migrating insect species, was the development of "barrier zone" methodology. A buffer zone of sterile screwworm releases was established to prevent infestations in the core eradication zone from immigrants outside the area. The extension of the "barrier zone" concept in the case of excluding establishment of PBW infestations in the San Joaquin Valley, CA, from migrating moths from the southwestern desert cotton growing areas appeared to be a logical approach.

C.   POTENTIAL APPLICATIONS FOR STERILE PBW MOTH RELEASES

Thus, the PBW sterile moth release method was envisioned as having two distinct and different potential applications: (1) suppression of established infestations in the southern desert valley cotton growing areas of California and Arizona, and (2) exclusion or prevention of infestation establishment in the PBW free San Joaquin Valley, CA.

II. SUPPRESSION OF ESTABLISHED PBW INFESTATIONS

A.   LITERATURE REVIEW

Field-cage tests in Texas with sterile PBW releases[6,7] and other field-cage studies,[12,13] reported significant reductions in population development of confined, known numbers of untreated PBW male and female moths in the presence of sterile moth releases with sterile to untreated moth ratios equal to or exceeding 50:1 (Table 1).

Field studies with sterile PBW moth releases for suppressing established PBW populations were only partially successful or failed. Bartlett[14] reported that results of 1966 unpublished data from tests conducted in partially isolated cotton fields in Arizona, showed approximately 64 percent reduction of PBW population development by release of sterile PBW

Table 1
Results of Field Cage Experiments to Evaluate Sterile
Pink Bollworm Moth Releases for Pink Bollworm Control

Source	Radiation dose (krad)	Sterile-to-native ratio generation		Control achieved as percent of check generation[a]	
		1	2	1	2
Ouye et al.[2]	[b]	[b]	[b]	[c]	[c]
Richmond and Graham[6]	25	50:1[d]	None	95[e]	91[e]
	40	50:1	None	87[e]	72[e]
Richmond and Graham[7]	15	25:1	55:1	67[e]	81[3]
	25	25:1	58:1	69[e]	85[e]
Bariola et al.[15]	10	20:1	[b]	21	[b]
	10	50:1	[b]	16	[b]
Flint et al.[12]	10	13-20:1	[b]	34	68[e]
	20	13-20:1	[b]	0	44
Flint et al.[13]	10	100:1	10:1	86[e]	[b]

[a] Total for season as calculated from author's data
[b] Not reported
[c] Suppression
[d] Single release at initiation of experiment
[e] Significantly different from control cage populations

moths obtained from 10-krad treated pupae. Sterile-to-native moth ratios were estimated to be 20:1, and the experiment was conducted over five PBW generations. A sterile PBW (25 krad) moth release experiment was conducted in a 2-ha cotton field at Borrego Springs, CA, in 1969.[16] Sterile to native moth ratios exceeding 10:1 were achieved only during 5 of 15 release weeks, whereas a 50:1 sterile:native ratio was assumed necessary to achieve population suppression. No measurable reduction of the native population occurred. A similar PBW sterile release program was conducted in the Coachella Valley, CA, in 1969 with similar results.[10] A second experiment was conducted at Borrego Springs in 1970 with sterile moths obtained from 10-krad irradiated pupae.[15] Again, effective overflooding ratios were not achieved, but 13 percent of 39 male larvae collected had chromosomal aberrations, whereas none were found in larvae collected in fields where no sterile moths were released. The results suggested native-sterile released moth mating interactions. Sterile PBW moth release studies were also conducted in 1971 and 1972 in the Moapa Valley, Nevada with 10-krad partially sterilized moths.[17] Results showed as great as a fivefold reduction of fertile progeny in release vs. nonrelease fields and significant decreases in moth populations in sterile release fields. Sterile (15 krad) PBW moth releases from 1968 to 1976 in wild cotton on the Florida Keys resulted in low infestation levels suggesting population suppression but infestations did not increase, as expected, following termination of the sterile moth releases.[10]

Lack of isolation from the influence of migrating native PBW populations into the experimental areas and/or high native populations in these studies made it difficult to obtain ratios of sterile-to-native insects sufficient to achieve population suppression. These factors, as well as possible adverse behavioral changes of the insects as a result of selection in mass-rearing that prevented mating interactions of released sterile with native insects,[18] were of concern. The need for further information defining the interaction of released sterile PBW moths with native moths under field conditions with established infestations led to studies on St. Croix, U.S. Virgin Islands.[19]

B.   ST. CROIX STUDIES
1.   Methods and Materials

St. Croix was picked as an experimental site because of: (1) the relative isolation of the island (about 64 km south of St. Thomas and St. John and 97 km southeast of Puerto Rico), (2) the absence of commercial cotton production eliminating conflicts with conventional production practices during evaluation of the program, and (3) the presence of an established PBW population. PBW infestations were first found in Sea Island cotton (*Gossypium barbardense* L.) bolls on St. Croix in 1921.[20] Populations have persisted in volunteer Sea Island cotton plantings since commercial production was terminated in 1927. Other PBW hosts on the island are wild cotton, *G.* spp., okra, *Hibiscus esculentis* L., *Thespesia populnea* L., and *Hibiscus vitifolius* L.[21] Surveys conducted from 1966 to 1968 on St. Croix indicated that PBW, as well as volunteer Sea Island cotton, were distributed over much of the island at elevations below 822 m.[22]

Sterile PBW moth releases were made by hand in cotton plots (*G. hirsutum* L.) maintained on the U.S. Department of Agriculture's Kingshill and nearby Virgin Islands Experiment Stations from January 1981 to April 1982. PBW moths were mass-reared and irradiated (20-krad gamma radiation, Co^{60}-irradiator) in the adult stage at the APHIS PBW rearing facility, Phoenix, AZ. Moth mortality and percentages of mated females during shipment and ability of the moths to mate after shipment were determined on St. Croix. Releases were made between the hours of 2 and 5 p.m. on the day of arrival. Released sterile moths contained Calco Red Oil dye, which was used to identify released moths from St. Croix moths.

Gossyplure-baited-Delta traps (Sandia Die and Cartridge Co., Albuquerque, NM) were used to study the seasonal distribution and abundance of St. Croix and sterile released PBW male moths. Blacklight Traps (BL)[23] were used to collect PBW female moths to determine the mating status of St. Croix and released sterile females. Both trapping methods were used to estimate sterile-release to St. Croix moth ratios.

Mating-table techniques,[24,25] moth hand collections in the sterile moth release plots,[26,27] and cytological examination[28] techniques were used to study sterile-St. Croix moth interactions.

Larval infestations in cotton bolls were determined as a measure of the impact of sterile moth releases on PBW population development.

2. Results

a. *Shipment Mortality, Mating Potential, and Released Sterile Moth Dispersal*

Mortalities of the irradiated, air-shipped, PBW moths, after 20 to 26, 40 to 56, and over 56 hours in transit were 7±5, 42±42, and 100 percent, respectively. Percentages of mated females in shipments arriving after 20 to 26 and 40 to 56 hours in transit were 6±5 percent and 5±5 percent, respectively; and 78 to 94 percent of moths of both groups in outdoor-insectary bioassays mated within 48 to 72 hours after arrival.

Sterile moths were released on 216 days during January 1, 1981 to April 1, 1982 (Figure 1). Average moth release was 103,061 per day but ranged from approximately 13 thousand to approximately 3 million for 1 to 13 release days per 2 week release period. The number of released sterile males caught in gossyplure baited traps were highly correlated to the number of traps operating and the number of released male moths.[19] Released sterile male moths were rarely captured more than 7 days after last releases (Figure 2). Over 40 percent of the total male moth captures occurred the first day after release. The number of released males captured thereafter decreased to less than 1 percent on day 7 following last releases. Most of the recaptured moths in gossyplure or BL traps occurred in or within 0.8 km of the release plots.

b. *Gossyplure-Baited Traps*

In 1980, St. Croix male moths were caught in gossyplure-baited traps every month of the year in the cotton plots (Figure 3). Peak numbers were caught in late March (20.4/trap/night) and late August (20.1/trap/night).

From January through April in 1981, after the initiation of sterile moth releases, St. Croix male moth catches ranged from 2.7 to 4.1/trap/night. But, with increasing ratios of released sterile males to St. Croix males caught

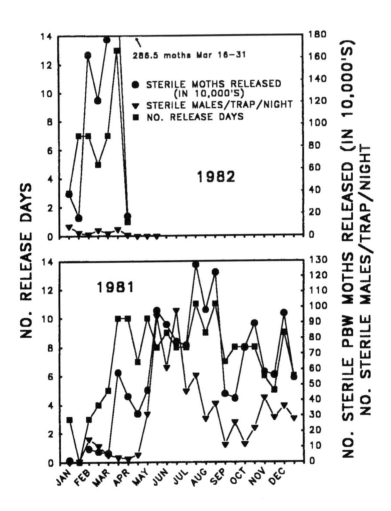

Figure 1. Number of release days, total number of sterile pink bollworm moths released per 2-week period, and number of sterile males caught per trap per night in gossyplure-baited traps.

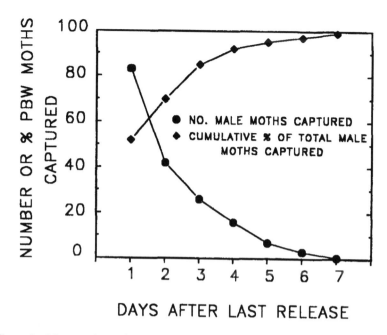

Figure 2. Mean number and cumulative percentages of sterile pink bollworm male moths captured in gossyplure-baited traps on days following releases.

in gossyplure-baited traps from May through December, numbers of St. Croix males caught ranged from 0.3 to 1.5/trap/night. When ratios of released sterile males to St. Croix males decreased during mid-February through March 1982, numbers of St. Croix males caught increased.

Average ratios of sterile to St. Croix males caught were low from January through April 1981, but averaged 72:1 from May 1 to August 31 and decreased to an average of about 21:1 from September to mid-October because of reduced numbers of sterile moth releases. Ratios of sterile to St. Croix males averaged 53:1 in November and December. Thereafter, ratios through March 31, 1982, averaged 11:1.

c. BL Traps

St. Croix male moths caught in BL traps in 1980 followed monthly trends similar to those reported for gossyplure-baited traps (Figure 4). BL trap catches of St. Croix males in January 1981 (sterile moth releases initiated December 29, 1980) averaged 3.6 males per trap per night. Numbers caught were low through mid-July, increasing thereafter to an average of 8.4 per trap per night in September, then decreasing to an average of 0.1 per trap per night from February 1 to 15, 1982 and increasing

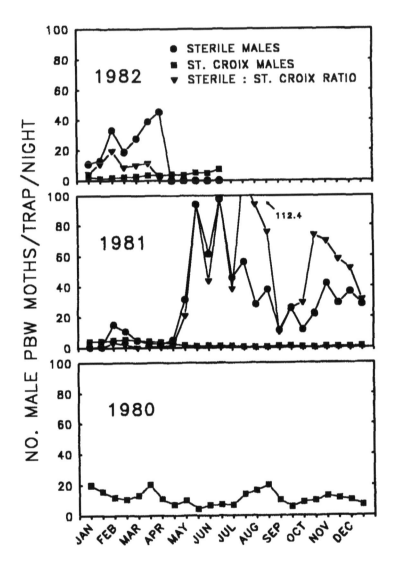

Figure 3. Mean number of pink bollworm sterile-released males, native St. Croix males, and sterile:native St. Croix male ratios per trap per night in gossyplure-baited traps.

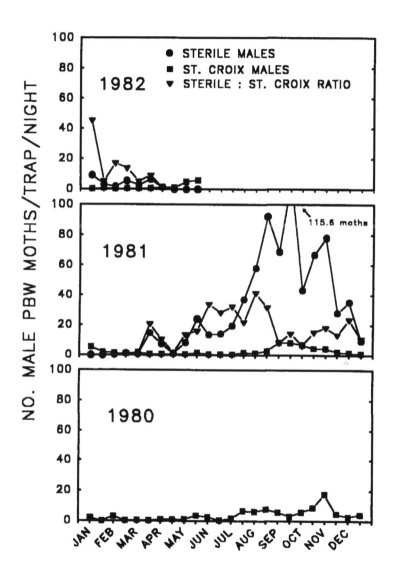

Figure 4. Mean number of pink bollworm sterile-released males, native St. Croix males, and sterile:native St. Croix male ratios per trap per night in blacklight traps.

following the last sterile moth releases. Captures of released sterile male moths in 1981 increased to a peak of 115.6 per trap per night during the September 16 to 30, 1981 sampling period. Numbers of released sterile male moths caught decreased thereafter through March 1982. Ratios of sterile-released to St. Croix males ranged from <1 to 46:1.

Numbers of St. Croix female moths caught averaged 1.8 per BL trap/night during 1980 (Figure 5). Percentages of mated St. Croix females ranged from 33 to 100 percent, and numbers of spermatophores per mated female ranged from 1.0 to 3.0. St. Croix and sterile female mating during the sterile moth releases (January 1981 to March 1982) ranged from 12 to 100 percent (1.2 to 2.4 spermatophores per female) and 12 to 60 percent (0.9 to 3.0 spermatophores per female), respectively.

St. Croix females caught in BL traps from January 1 through April 15, 1981, contained no red spermatophores from released sterile male moths (Figure 6). From April 16 to 30, to October 1 to 15, 1981, St. Croix females were found that had mated with released sterile males during 10 of the 12 sampling periods. Percentages of St. Croix females with red spermatophores during those 10 sampling periods ranged from 17 to 50 percent.

Observed mean numbers of (red-dyed spermatophores), January 1 to October 15, released sterile males mating with St. Croix females, were significantly greater than expected based on gossyplure-baited male trap ratios but no different than expected based on BL trap sterile male ratios (Table 2). St. Croix male to female mating using either gossyplure-baited or BL Trap ratios was significantly greater than expected.

From October 16, 1981, to March 31, 1982, red and white spermatophores were identified and recorded from dissections of both released sterile and St. Croix females caught in BL traps (Figure 7). The highest percent (57 percent, red spermatophores) of collected St. Croix female to released sterile male mating occurred during the November 16 to 30, 1981, sampling period when ratios of sterile to St. Croix males averaged more than 50 to 1. St. Croix female-male mating ranged from 43 to 100% of the numbers counted.

Using gossyplure-baited trap ratios (sterile male to St. Croix male), observed matings of released sterile males to St. Croix females and of St. Croix males to released sterile females were greater than expected (Table 2). Released sterile male to released sterile female matings were less than expected, and St. Croix male to St. Croix female matings were significantly greater than expected. For the entire sampling period, matings of released sterile males to St. Croix females, of released sterile males to released sterile females, and of St. Croix males to released sterile females were not different than expected on the basis of BL trap sterile male to St. Croix male ratios. St. Croix male to St. Croix female matings were greater than expected.

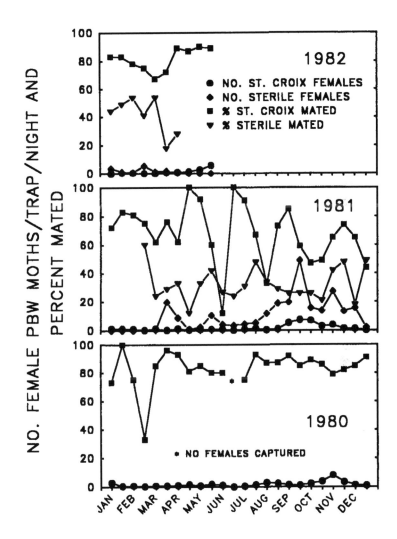

Figure 5. Mean number and percent mated pink bollworm sterile-released female and native St. Croix female moths captured in blacklight traps.

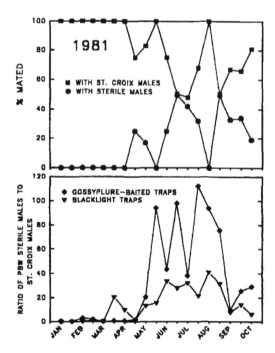

Figure 6. Mean pink bollworm sterile released male to St. Croix male moth ratios and mean percentages of St. Croix females mated with sterile or St. Croix males (based on red-dyed spermatophores).

d.   Mating Tables

There were no significant differences between the percentages of clipped-wing irradiated, untreated mass-reared, and St. Croix females mating with St. Croix males on mating tables (Table 3).

e.   Hand Collections: Sterile-Native Moths

All combinations of mating pairs were collected in cotton plots during sterile-moth releases (Table 4). Based on numbers of sterile and native moths collected, the numbers of mating pairs of released sterile males and St. Croix females and of released sterile females and St. Croix males were less than expected. Collected numbers of mating pairs of released sterile males and females and of St. Croix males and females were greater than expected.

f.   PBW Larval Infestations in Cotton Bolls

In 1980, boll infestations in cotton plots ranged from 1.2 to 3.8 larvae per boll from January through July, peaked at 7.6 larvae per boll during the August 16 to 31 sampling period, and decreased thereafter to 1.4 larvae per boll in late December.

Table 2

Numbers of Expected and Observed Released Sterile (S) and St. Croix (N) Female Pink Bollworm Moths Mating with Released Sterile or St. Croix Male Moths[a]

Sampling periods and mating pair combinations[b] ♂	♀	Expected	Observed	χ^2	P
Jan. 1 to Oct. 15, 1981		Gossyplure trap ratios[c]			
S	N	45	114	100.0	<0.01
N	N	1	405	400.0	<0.01
		BL trap ratios			
S	N	90	114	0.2	<0.50
N	N	6	405	400.0	<0.01
Oct. 16, 1981 to Mar. 31, 1982					
		Gossyplure trap ratios[c]			
S	N	34	86	7.1	<0.01
S	S	1078	821	5.6	<0.01
N	S	34	109	14.9	<0.01
N	N	1	130	100.0	<0.01
		BL trap ratios			
S	N	69	86	0.4	<0.50
S	S	1009	821	3.3	<0.10
N	S	69	109	2.1	<0.10
N	N	5	130	>100.0	<0.01

[a] Based on 519 native St. Croix females from January 1 to October 15, 1981 and 216 native and 930 sterile blacklight trapped pink bollworm female moths captured from October 16, 1981 to March 31, 1982. Sterile released male matings distinguished by the presence of calco red dyed spermatophores.

[b] S = Sterile, N = native St. Croix.

[c] Sterile female - native female ratios assumed to be the same.

In 1981, larval infestations were 1.2 larvae per boll when sterile moth releases were initiated. Infestations averaged 0.8 to 1.2 larvae per boll through May 1 to 15. Beginning May 16 to 31, infestations dropped to 0.5 larvae per boll, decreased to 0.3 to 0.4 larvae per boll in August, increased thereafter to 1.5 larvae per boll September 1 to 15, and ranged from 1.1

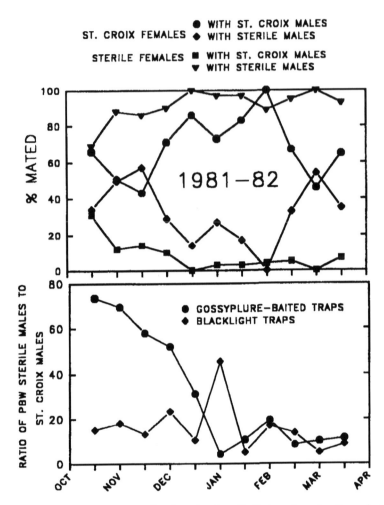

Figure 7. Mean pink bollworm sterile released male to St. Croix male moth ratios and mean percentages of St. Croix and sterile females mated with St. Croix or sterile males (based on red-dyed spermatophores).

to 2.9 larvae per boll from the September 16 to 30 sampling period through the end of December 1981. Infestations ranged thereafter from 0.9 to 1.8 larvae per boll during January 1 to 15, 1982, to April 1 to 15, 1982, and increased after the last release of sterile moths to 3.7 larvae per boll (Figure 8).

Further substantiation of the mating interaction of the sterile release moths with St. Croix moths was suggested with the identification of chromosomal aberrations in 7, 10, 15, 0, 4, and 0 percent (119 total larvae) of PBW

Table 3

Mean Mating Percentages of Clipped-wing-mass Reared Untreated and Irradiated and Untreated St. Croix, Virgin Female Moths on Mating Tables During Pink Bollworm Moth Sterile Releases[a]

Moth Strain	Number	Percent Mated	Percent mated with[b] St. Croix ♂	Released sterile ♂
Mass Reared				
Untreated	1252	66 a	64 a	36 a
Irradiated (20kr)	1376	50 a	58 a	42 a
St. Croix	322	47 a	53 a	47 a

[a] 20 to 30 virgin females on each of two to four mating tables per night. Total of 36 nights from 8 Jan. 1981 to 19 March 1982. Means in a column followed by the same letter are not significantly different.

[b] Determined on the basis of the presence of a red-dyed spermatophore for sterile male mating.

Table 4

Expected and Observed[a] Sterile-released (S) and St. Croix (N) Combinations of Mating Pairs Hand Collected in Cotton Plots During Release of Sterile Pink Bollworm Moths

Mating Combinations Male	Female	Exp.	Obs.	χ^2	P
S	N	73	12	50.9	<0.01
S	S	388	463	14.5	<0.01
N	S	47	24	11.3	<0.01
N	N	10	19	8.1	<0.01

[a] Based on ratios of total numbers of St. Croix (N) and released sterile (S) moths hand collected in cotton plots on 29 nights during January 15, 1981 to March 25, 1982.

larvae collected on October 7 and November 12, 1981; February 5 and 23, and May 6 and 10, 1982, respectively).

3. Summary

Sterile PBW moth releases on established field infestations under the fairly isolated conditions of the reported studies required high numbers (over 100,000 per release day per 0.5 ha) of released sterile PBW moths to achieve suppression of larval infestations in bolls. These results, supported by analyses of release-sterile-native PBW moth interactions, suggest the non-

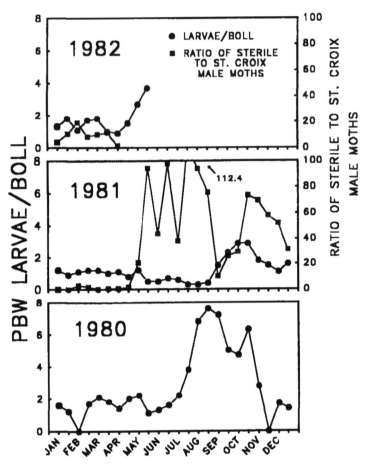

Figure 8. Mean number of pink bollworm larvae per cotton boll before (1980), during (1981 to March 1982), and following (April to June 1982) sterile moth releases with ratios of sterile-released to native St. Croix male moths captured in gossyplure-baited traps.

competitive nature of the mass-reared released PBW moths. The effect of the handling and shipping procedures on the overall vigor of the released, sterile PBW is difficult to assess. Mortality of the released insects was apparently much higher than occurred in laboratory quality control monitoring since few or no released sterile males were caught in baited traps more than seven days after last releases. Thus, continuity and frequency of releases is an important consideration in maintaining effective overflooding sterile to native moth ratios. Also, dispersal of sterile released PBW moths under the experimental conditions was less than expected, which necessitates concern for the need for distribution of the sterile released PBW over the cotton ecosystem in suppression programs.

Mass-reared virgin females attracted and mated competitively on mating tables with clipped-wing, St. Croix females for released sterile or St. Croix males. The contribution of released sterile PBW females in release programs has not been extensively studied but should be evaluated in future research. The interaction of released sterile moths with St. Croix moths in the population was substantiated by the relatively high (up to 15 percent) percentages of male larvae from cotton bolls found with chromosomal aberrations. Hand collections of mating pairs in release plots were the most direct evidence of the mating interactions of released sterile and St. Croix insects. However, the fact that fewer released sterile male to St. Croix female and released sterile female to St. Croix male mating pairs were found than expected and more St. Croix male to female mating pairs were collected than expected is a further indication of the noncompetitive nature of the sterilized, mass-reared, released PBW moths. Expected and observed mating interactions of sterile released and St. Croix moths based on gossyplure and BL trap catch ratios were variable and conflicted with these results and remain unexplained, although trap efficiency, overloading phenomena, and other factors affecting trap variability may be involved.

Reduced boll infestations occurred when ratios of released sterile to St. Croix males averaged 72:1. Infestations increased when the sterile to St. Croix male ratios averaged 20:1. But the 20:1 released sterile to St. Croix male ratio apparently had some population suppression effect since larval infestations increased from about one larva per boll to 3.7 larvae per boll during the 2 months immediately after the last release of sterile insects.

The low-level competitive nature of the mass-reared sterile release PBW moths necessitating higher than theoretical ratios of sterile to release insects to achieve population suppression under established infestation field conditions leaves ample room for improvement. The biological significance:" (1) of reduced ability of sterile male PBW moths to effectively inseminate females,[29] (2) to transfer apyrene sperm (sterile-released males or their progeny),[30] and (3) the effects on reproductive behavior of involved females have not been defined in cage or field tests but may affect the efficiency of sterile releases for control or suppression of PBW populations.[31,32] These issues would be productive research areas that could provide knowledge useful in improving the system. Other concerns involve the effect of colonization and laboratory-rearing procedures that may have resulted in gradual selection of less vigorous and behaviorally different strains affecting field performance of released insects interacting with native populations. Genetic selection in mass-rearing of a trait(s) that physically, physiologically, or behaviorally prevents males from responding to native females (or the native females become nonreceptive as a result of altered physical, physiological or behavioral patterns of the released, mass-reared sterile males) could have a significant effect on the impact of sterile PBW release programs. These issues have not been resolved, but should be fully investigated with a view to the production of insects that are as close to normal vigor and competitiveness as possible.

III. STERILE PBW MOTH RELEASES TO EXCLUDE OR PREVENT ESTABLISHMENT IN THE SAN JOAQUIN VALLEY, CA

A. PBW MOTH MIGRATION

The demonstration of PBW moth flight and its role in the spread and establishment of cotton infestations from Mexico to Texas,[33-35] and the dispersal of the PBW under arid desert conditions as shown by Bariola et al.[36] and Stern[37] supported the cotton industries concern regarding the potential for moth movement into the San Joaquin Valley from southern desert areas. Additionally, Kauper[38] showed that favorable windflows occurred when low pressure areas developed off the southern California coast for approximately 2 days. Favorable wind movement for PBW moth dispersal occurred 10 times during a selected 13-month sampling period and provided 25 potential moth migration days.[39]

B. PBW STERILE MOTH RELEASE INTEGRATED MANAGEMENT PROGRAM

The emergency areawide sterile PBW moth release program in the San Joaquin Valley was initiated in 1968. The program, over the years, has evolved technologically to currently consist of: (1) PBW traps baited with gossyplure to detect native migrant moths and indicate areas of needed suppressive action as well as to establish ratios of released sterile to native male moths, (2) release of radiation-sterilized moths, (3) most recently, behavioral control involving field application of gossyplure slow release systems, and (4) cotton plant destruction and plow down to maintain a 90-day host-free period.[40]

1. Native Moth Detection and Sterile Moth Release Monitoring

A vital complement to the sterile insect release system was methodology for: (1) PBW detection at low population densities and (2) estimating the ratio of sterile insects to native PBW as a measure of the potential impact of the sterile-release system on the native population. This essential component of the program was provided with the identification and synthesis of the PBW sex pheromone, gossyplure (1:1 ratio of Z,Z and Z,E isomers of 7,11-hexadecadienyl acetate).[41] When gossyplure was incorporated into the system and the Fricke trap was replaced with the Delta trap[42] in 1974, a dramatic increase in efficiency of the detection and monitoring phases of the sterile-release program occurred.[10] Numbers of native moths caught from 1974 through 1990 have ranged from a low of 69 in 1978 to a high of 7402 in 1977 as compared to 0 to 25 PBW detected from 1968 to 1973 using other trapping systems.

2. Sterile Moth Releases

In 1968 and 1969, PBW were reared and irradiated (40 krad) at the USDA, ARS laboratory at Brownsville, Texas. Sterilized insects were subsequently shipped to Bakersfield, California for release.[43] The 1968 efforts were focused on approximately 5871 ha of cotton surrounding the four locations of the 1967 native PBW moth and infested cotton boll collection sites. No native moths were collected in 1968 in the area nor elsewhere in the San Joaquin Valley. However, sterile moth releases were made in 1969 in the original moth find areas of 1967. Also, five native moths were trapped in 1969 in an area 4.4 km southwest of Bakersfield and the sterile moth releases increased to include this area in the 1970 program. Since 1970, mass-rearing of PBW moths for sterilization and release has been accomplished at the USDA, Animal and Plant Health Inspection Service (APHIS) laboratory at Phoenix, Arizona, except for several years in the late 70's, when commercial contractors supplied one million moths per day to supplement APHIS production. In 1970, larvae infesting bolls and native moth catches in new areas necessitated the inclusion of additional cotton acreage requiring sterile moth releases. Similar patterns of increasing and decreasing acreage on which sterile moth releases have been made have occurred in ensuing years. A number of factors probably influence this pattern of activity including PBW moth dispersion, intensiveness of trapping activity, and increasing levels of experience with identified areas of traditional occurrence of native PBW (native moths and/or larval infestations).

In 1968, nine million sterile PBW moths were released. The numbers released have increased consistently in subsequent years, except for 1972 to 1974, when disease problems in rearing were encountered. In 1991 over 800 million sterile insects were released over 142 thousand ha of cotton in the valley. Annual ratios of sterile released to native moths as measured by captures in gossyplure-baited traps have ranged from about 100:1 to 6200:1.[10] Native male moths have been trapped in the San Joaquin each year of the program since 1969 and larvae found in bolls in each of 6 years. Diapausing PBW larvae have been demonstrated to survive, pupate, and emerge in the spring in the Bakersfield area of the San Joaquin Valley.[44]

3. Behavioral Control and Incipient Infestations

The low level, localized larval boll infestations and occasional multiple PBW moth catches in localized areas has become of increasing concern in the San Joaquin Valley. The behavioral control approach using gossyplure has been particularly appealing in the San Joaquin Valley Program because it is a complete biological system when the pheromone alone is used and includes only a minimal amount of insecticide when the pheromone and low-level insecticide formulation is used to attract and kill male moths.[45-51]

Behavioral control methodology using the pheromone alone or pheromone plus low-levels of insecticide as supplementary suppression control measures in the San Joaquin Valley have been employed when: (1) 20 or more native moths are caught in a one-mile section (prior to October), (2) there is evidence of a reproducing generation, (3) larvae are found, or (4) the native to sterile moth ratio is less than 1:50.[40] The behavioral control component of the sterile PBW management system has been implemented during six of the growing seasons from 1983 to 1990 (California Department of Food and Agriculture, Sacramento, CA, reports for 1983-1990). In 1983, the pheromone slow release formulation was applied after being triggered by finding larval infestations in bolls and in 1984, 1985, 1987, 1988, and 1990 by early-season moth trap catches suggesting overwintering populations or by successive moth trap catches at intervals that suggested generation cycling and reproduction. Early harvest and early crop destruction and plowdown have been accomplished in areas where the necessity for behavioral control treatments have been indicated. Also, in each year following the treatment year, areas treated were considered "high hazard" for program purposes and gossyplure-baited trappings were initiated one month earlier than under regular trapping schedules to detect overwintering populations. In all cases, populations in subsequent years have been reduced to nondetectable levels, suggesting that the combined action of behavioral control, sterile moth releases, and the host-free period were effective in eliminating actual or suspected incipient infestations.

In 1990, high moth trap catches triggered 4-6 pheromone treatments in 14 fields in the Valley, involving about 680 ha of cotton.[52] Further, the highest larval finds in cotton bolls in the San Joaquin Valley occurred since the introduction of the pest in California. A 250-boll sample in one field yielded 38 PBW larvae, and a second field had 2 larvae/250 bolls. Also, several bolls were found with exit holes. Infested fields were insecticide treated, harvest dates moved forward and shredding, disking, deep plowing and winter irrigation practices accomplished. The infested area was extensively trapped in the 1991 season and subjected to early releases of high numbers of sterile moths. The results of these efforts were apparently successful since no dramatic increase in the PBW population occurred in 1991.

4.   Host-Free Period

The beneficial contribution of a host-free period in PBW management has long been established.[53] Although it is difficult to quantify the value of this cultural practice in the San Joaquin Valley program, the benefits have been well documented in areas where the PBW is an established annual pest. Stalk shredding to enhance uniform and deep burial of shredded plant debris, followed by disking and effective plowing and winter irrigation treatments, effectively reduces the numbers of overwintering PBW. The earlier winter plowing is accomplished, the higher the larval mortality, with fewer moths emerging in the spring. The 90-day host-free period maintained in the San Joaquin Valley is a vital component in the PBW management system. Shortening the growing season with efficient water management schedules, early defoliation and/or plant growth regulator treatment, early harvest, and crop destruction deprive the PBW of the opportunity to develop a diapause generation and minimizes the probability of overwintering.

5.   Summary

The integrated management program in the San Joaquin Valley has been in effect for more than 20 years. Thus, irrespective of annual PBW moth migration into the valley, and the proven ability of the PBW to overwinter in the area, established infestations have not occurred. Costs for the program through 1990 have not exceeded $4 per acre and no sustainable cotton production loss attributable to infestations has occurred.[40]   This, compared to a conservative estimate of $75 to $100 control costs per acre in the lower desert cotton growing areas of California, plus yield losses[54] and declining cotton acreage in some areas, as well as the threat of secondary pests and increasing levels of PBW and secondary pest insecticide resistance suggest that the integrated PBW management program initiated in 1968 has been of great economic and ecological benefit to the San Joaquin Valley agricultural communities.   The only way to conclusively prove the contribution of the sterile PBW moth releases to the program success would be to terminate the sterile moth releases followed by comparisons of PBW infestation trends and native moths with those observed during the years of the release program.   In view of the risk involved, a more logical and reasonable approach appears to be continued research effort to analyze and evaluate the insects released, their interaction in the population, to undertake suitable pilot programs to determine their impact of migrating PBW populations, as well as on low-level established field infestations, and implement new population suppression components into the system as they become available.

IV. DISCUSSION

A.   SUPPRESSION OF ESTABLISHED PBW INFESTATIONS

The results of laboratory and field cage studies have suggested potential for the use of sterile PBW moth releases for suppression of established PBW infestations.   Under experimental field or commercial cultivated cotton growing conditions, demonstration of the impact of PBW sterile moth releases has been much more difficult to achieve.   However, the St. Croix results and recent large area demonstration trials suggest that sterile PBW moth releases in combination with behavioral control methodology may be an effective PBW management system for low level established infestations.[55,56]

B.   STERILE PBW MOTH RELEASES TO EXCLUDE OR PREVENT ESTABLISHMENT IN THE SAN JOAQUIN VALLEY, CALIFORNIA

The results of research designed to evaluate the effect of sterile PBW moth releases on established infestations are not directly applicable to the San Joaquin program.   Ecological differences in weather parameters affecting population development and PBW behavior preclude direct comparisons. Also, no measurable PBW population occurs in the San Joaquin Valley and ratios of released sterile moths are excessively high in comparison to any ratios that have occurred in experimental trials.   The focus of the program is to provide overwhelming numbers of sterile PBW males and females in the ecosystem to absorb the reproductive potential of migrant male and virgin migrant female moths.   The impact of fertilized migrant females is virtually unknown.   But if reproduction occurs, low levels of resulting progeny should be exposed to the extremely high ratios of released sterile PBW moths resulting in low probability of reproduction occurring.   The specialized use of sterile PBW moth releases in the San Joaquin Valley has also been integrated into a system of efficient detection, monitoring, behavioral control, and host-free period management.

The gossyplure trap monitoring system provides information on male PBW migrants only.   The number of migrating female moths and their age and mating status is unknown, but is important information necessary to assess the reproductive potential of the migrating population.   Larval infestations have on occasion occurred in the San Joaquin Valley, but have not persisted because immediate and thorough suppressive actions have been taken.   The large numbers of sterile insects released provide the potential for some reproduction as a result of released moths mating. F_1 sterile males would be unmarked male moths indistinguishable from migrant moths.   This issue has been of continuing concern during the conduct of the program. Also, F_1 female progeny of irradiated (20 krad) moths are not completely sterile and pose an additional possibility to explain limited numbers of native insect finds.[5,32]   The increasing frequency of larval infestations clearly indicates that some level of reproduction has occurred in cotton in the San Joaquin Valley.   Release of the "sooty" genetic marker strain[57] has been considered to resolve this issue, but preliminary trials have been unsuccessful.

A solution to this problem remains critical. Reproduction, even at a very low level, in a small percentage of the released insects could result in serious misinterpretation of the origin of pheromone-baited native male moth trap captures.

Behavioral control remains a highly desirable option to supplement sterile releases in the San Joaquin Valley PBW management system when low level incipient infestations occur. The most recent gossyplure slow release system was developed by Shin-Etsu Chemical Industry, Co., LTD, Tokyo, Japan. It is a hollow polyethylene tube containing about 78 mg of gossyplure and called PBW-ROPE®. Promising PBW control was obtained using PBW-ROPE, manually tied to cotton plants, in numbers to result in 64 g AI/ha of gossyplure.[55,58] Subsequently, large scale field trials involving 30 cotton fields in Coachella Valley, CA, resulted in extremely low cotton boll infestations.[56] The second year of the field trial, sterile PBW moths were released and PBW-ROPE was applied only to fields where ratios of sterile to native males in gossyplure-baited traps were less than 60:1. Boll infestations were kept below economic levels and only 6 of 27 fields required insecticide treatments. Thus the large-scale demonstration trials suggest another available technology for implementation in the areawide management program in the future.

C.   GENERAL

The ultimate solution to the PBW problem in the southwest cotton growing areas is the development of an areawide suppression program focusing on the PBW population in areas where the insect is currently established. The technology for achieving a high degree of suppression of established native populations is well advanced,[59] and may be well within the range of practical feasibility.

Research, extension, and teaching efforts dealing with the PBW have made significant progress in development of the concept of coordinated large area, agricultural community involvement in managing the pest. The areawide approach focuses suppressive measures on the total PBW population as opposed to uncoordinated efforts focused on local or farm-to-farm or field-by-field attempts to control limited segments of the PBW population. The most effective and efficient areawide PBW management programs incorporate multifaceted, multidiscipline inputs to achieve population suppression with little or minimal impact on other components of the environment. Chemical, biological, behavioral, genetic, and cultural control methods, as well as development of resistant cotton variety technology, is advancing rapidly. All control methods must be considered in areawide PBW suppression systems. No single method is totally acceptable, and combinations of two or more methods offer the highest probability of success in suppression/management programs. The selection and integration of compatible control methods must be based on knowledge of how each of the methods function individually and when introduced separately or simultaneously as suppression methods to achieve population reduction. Further, suppressive action must be taken within the framework of detailed knowledge of the biology, ecology, and population dynamics of the target species as well as crop development. The potential long-term benefits of PBW population suppression on an areawide basis in areas where it is an established annual economic problem appear to justify the efforts in terms of reduced costs, more effective control, less environmental contamination, and other peripheral problems associated with local uncoordinated efforts which have not influenced the occurrence of year-after-year economic pest populations.

The potential use of a sterile PBW moth integrated management system for suppression of established populations in southwestern cotton-producing areas is a distinct possibility if indigenous PBW numbers can be reduced to low levels using a combination of cultural and other control methods. PBW mass rearing capability has improved dramatically in recent years. Increased efforts to mechanize many of the steps in the rearing protocol will undoubtedly further increase efficiency and production at reduced costs.

ACKNOWLEDGMENT

The author appreciates the information provided by Len Foote, California Department of Food and Agriculture, Sacramento, CA, concerning the San Joaquin Valley, CA, pink bollworm sterile release program.

REFERENCES

1.   **Knipling, E. F.,** Screwworm eradication: concepts and research leading to the sterile male method, in *Smithsonian Report for 1958*, Smithsonian Publication 4365, 1959, 409.

2.   **Ouye, M. T., Gracia, R. S., and Martin, D. F.,** Determination of the optimum dosage for pink bollworm treated as pupae with gamma radiation, *J. Econ. Entomol.*, 57, 387, 1964.

3.   **Vanderzant, E. S. and Reiser, R.,** Aseptic rearing of the pink bollworm on synthetic diet, *J. Econ. Entomol.*, 49, 7, 1956.

4.   **Richmond, C. A. and Ignoffo, C.,** Mass rearing pink bollworms, *J. Econ. Entomol.*, 57, 503, 1964.

5.   **Graham, H.,** Dosages of gamma irradiation for full and inherited sterility in adult pink bollworms, *J. Econ. Entomol.*, 65, 645, 1972.

6.   **Richmond, C. A. and Graham, H. M.,** Suppression of populations of pink bollworms with releases of sterilized moths in field cages, *J. Econ. Entomol.*, 63, 1366, 1970.

7.   **Richmond, C. A. and Graham, H. M.,** Suppression of populations of pink bollworm by releases of gamma irradiated moths in field cages, *J. Econ. Entomol.*, 69, 332, 1971.

8.   **Harper, R. W.,** Pink bollworm of cotton in California, *California Dept. Food and Agric.,* Sacramento, CA, 1967.

9.   **Harper, R. W. and Gammon, C.,** Pink bollworm-boll weevil progress report 66-2, *California Dept. Food and Agric.,* Sacramento, CA, 1966.

10.   **U.S. Department of Agriculture,** Task force review of the pink bollworm program. *USDA, APHIS, Plant Protection and Quarantine,* 1977.

11.   **Baumhover, A. H.,** Eradication of the screwworm fly, an agent of myiasis, *J. Amer. Med. Assoc.,* 196, 240, 1966.

12.   **Flint, H. M., Palmer, D. L., Bariola, L. A., and Horn, B.,** Suppression of populations of native pink bollworm in field cages by release of irradiated moths, *J. Econ. Entomol.,* 67, 55, 1974.

13.   **Flint, H. M., Wright, B., Sallam, H. A., and Horn, B.,** A comparison of irradiated or chemosterilized pink bollworm moths for suppressing native populations in field cages, *Can. Entomol.,* 107, 1069, 1975.

14.   **Bartlett, A. C.,** Radiation induced sterility in the pink bollworm, *USDA, ARS, ARM-W-1,* 1978.

15.   **Bariola, L. A., Bartlett, A. C., Staten, R. T., Rosander, R. W., and Keller, J. C.,** Partially sterilized adult pink bollworms: releases in cages and field cause chromosomal aberrations, *Environ. Entomol.,* 2, 173, 1973.

16.   **Graham, H.,** Sterile pink bollworm: field releases for population suppression, *J. Econ. Entomol.,* 71, 233, 1978.

17.   **Staten, R., Brazzel, J. R., Bartlett, A. C., and Robison, G. D.,** Suppression of a native moth pink bollworm population through the release of mass-reared partially sterilized moths, APHIS, unpublished report, 1973.

18.   **Boller, E.,** Behavioral aspects of mass rearing insects, *Entomophaga,* 17, 9, 1972.

19.   **Henneberry, T. J. and Keaveny III, D. F.,** Suppression of pink bollworm by sterile moth releases, *USDA, ARS, ARS-32,* National Tech. Info. Service, Springfield, VA, 1985.

20.   **Smith, L.,** Report of the Virgin Islands Agricultural Experiment Station, 1921, *Virgin Islands Agric. Exp. Stn.,* 1922.

21.   **Loftin, U. C.,** Summary of work on the pink bollworm. Report of the Virgin Islands Agricultural Experiment Station, *Virgin Islands Agric. Exp. Stn.,* 1932.

22.   **Graham, H. and Cantelo, W. W.,** Populations of the pink bollworm on St. Croix, U.S. Virgin Islands, *J. Econ. Entomol.,* 66, 266, 1973.

23.   **Harding, W. C. Jr., Hartsock, J. G., and Rohwer, G. G.,** Black light trap standards for general insect survey, *Bull. Entomol. Soc. Am.,* 12, 31, 1966.

24.   **Snow, J. W., Raulston, J. R., and Guillot, F. S.,** Mating tables: a method to study the mating and behavior of lepidoptera and diptera under field conditions, *Ann. Entomol. Soc. Am.,* 69, 751, 1976.

25.   **Lingren, P. D., Raulston, J. R., Sparks, A. N., and Proshold, F. I.,** Tobacco budworm: nocturnal behavior of laboratory-reared, irradiated and native adults in the field, *USDA, SEA, ARR-W-5,* 1979.

26.   **Raulston, J. R., Graham, H. M., Lingren, P. D., and Snow, J. W.,** Mating interaction of native and laboratory-reared tobacco budworms released in the field, *Environ. Entomol.,* 5, 195, 1976.

27.   **Lingren, P. D., Raulston, J. R., Sparks, A. N., and Wolf, W. W.,** Insect monitoring technology for evaluation of suppression via pheromone systems, in *Insect Suppression with Controlled Release Pheromone Systems,* Kydonieus, A. F. and Beroza, M., Eds., CRC Press, Boca Raton, FL, 1982, 171.

28.   **Bartlett, A. C. and Lewis, L. J.,** Pink bollworm: chromosomal damage and reproduction after gamma irradiation of larvae, *J. Econ. Entomol.,* 66, 731, 1973.

29.   **Cheng, W. and North, D. T.,** Inherited sterility in the F$_1$ progeny of irradiated male pink bollworms, *J. Econ. Entomol.*, 65, 1272, 1972.

30.   **LeChance, L. E., Richard, R. D., and Proshold, F. I.,** Radiation response in pink bollworm: a comparative study of sperm bundle production, sperm transfer, and oviposition response elicited by native and laboratory-reared males, *Environ. Entomol.*, 4, 321, 1975.

31.   **Henneberry, T. J. and Clayton, T.,** Pink bollworm: mating, reproduction, and longevity of laboratory-reared and native strains, *Ann. Entomol. Soc. Am.*, 73, 382, 1980.

32.   **Henneberry, T. J. and Clayton, T.,** Effects on reproduction of gamma irradiated, laboratory-reared pink bollworms and their F$_1$ progeny after matings with untreated laboratory-reared or native insects, *J. Econ. Entomol.*, 74, 19, 1981.

33.   **Coad, B. R.,** Organization and progress of pink bollworm research investigations, *J. Econ. Entomol.*, 22, 743, 1929.

34.   **Fenton, F. A.,** Biological notes on the pink bollworm [*Pectinophora gossypiella* (Saunders)] in Texas, in *Proc. IV International Congress of Entomol.*, 1929, 439.

35.   **McDonald, R. E. and Loftin, U. C.,** Dispersal of the pink bollworm by flight or wind carriage of the moths, *J. Econ. Entomol.*, 18, 745, 1935.

36.   **Bariola, L. A., Keller, J. C., Turley, D. L., and Farris, J. R.,** Migration and population studies of the pink bollworm in the arid West, *Environ. Entomol.*, 2, 205, 1973.

37.   **Stern, V. M.,** Long and short range dispersal of the pink bollworm *Pectinophora gossypiella* over southern California, *Environ. Entomol.*, 8, 524, 1979.

38.   **Kauper, E. K.,** Air flow analysis - Imperial Valley to San Joaquin Valley, unpublished report, 1977.

39.   **Wolf, W. W. and Kauper, E. K.,** A mechanism for aerial transport of pink bollworms to the San Joaquin Valley of California, unpublished report, 1980.

40.   **Foote, L.,** Pink bollworm program in the San Joaquin Valley, California, in *Proc. International Cotton Pest Work Committee*, San Jose del Cabo, B.C. Mexico, 1989, 7.

41.   **Hummel, H. E., Gaston, L. K., Shorey, H. H., Kaae, R. S., Byrne, K. S., and Silverstein, R. M.,** Clarification of the chemical status of the pink bollworm sex pheromones, *Science*, 181, 873, 1973.

42.   **Foster, R. N., Staten, R. T., and Miller, E.,** Evaluation of traps for pink bollworm, *J. Econ. Entomol.*, 70, 289, 1977.

43.   **Forbes, A. G. and Reed, B. C.,** Report to the 1972 Legislature on the pink bollworm control program pursuant to 1972 assembly concurrent resolution No. 38 relative to the pink bollworm of cotton program, State of California Agriculture and Services Agency, Depart. of Agric., Division of Plant Industry, Control and Eradication, Sacramento, CA, 1972.

44.   **Bartlett, A. C. and Staten, R. T.,** unpublished data.

45.   **Gaston, L. K., Kaae, R. S., Shorey, H. H., and Sellers, D.,** Controlling the pink bollworm by disrupting sex pheromone communication between adult moths, *Science*, 196, 904, 1977.

46.   **Brooks, T. W., Doane, C. C., and Staten, R. T.,** Experience with the first commercial pheromone communication disruptive for suppression of an agricultural pest, in *Chemical Ecology: Odour Communication in Animals*, Ritter, F. J., Ed., Elsevier, Amsterdam, 1979, 375.

47.   **Doane, C. C. and Brooks, T. W.,** Research and development of pheromones for insect control with emphasis on the pink bollworm, in *Management of Insect Pests with Semiochemicals: Concepts and Practice*, Mitchell, E. R., Ed., Plenum, New York, 1981, 285.

48. Henneberry, T. J., Gillespie, J. M., Bariola, L. A., Flint, H. M., Lingren, P. D., and Kydonieus, A. F., Gossyplure in laminated plastic formulations for mating disruption and pink bollworm control, *J. Econ. Entomol.*, 74, 376, 1981.

49. Butler, G. D., Jr., Henneberry, T. J., and Barker, R. J., Pink bollworm: comparison of commercial control with gossyplure or insecticides, *USDA, ARS, ARM-W-35*, 1983.

50. Critchley, B. R., Campion, P. G., McVeigh, L. J., Hunter-Jones, P., Hall, D. R., Cork, A., Nesbitt, B. F., Marrs, G. J., Jutsum, A. R., Hosney, M. M., and Nasr, E. A., Control of pink bollworm, *Pectinophora gossypiella* (Saunders) (Lepidoptera: Gelechiidae), in Egypt by mating disruption using aerially applied microencapsulated pheromone formulation, *Bull. Entomol. Res.*, 73, 289, 1983.

51. Beasley, C. A. and Henneberry, T. J., Pink bollworm-gossyplure studies in Palo Verde Valley, California, *USDA, ARS, ARS-39*, National Tech. Info. Service, Springfield, VA, 1985.

52. Goodell, P. B. and Bently, W., Pink bollworm in the San Joaquin Valley - 1990 update, *California Cotton Rev.*, 18, 4, 1990.

53. Watson, T. F., Methods for reducing winter survival of the pink bollworm, in *Pink Bollworm Control in the Western United States*, H. Graham, Ed., USDA, SEA, ARM-W-16, 24, 1980.

54. Burrows, T. M., Sevacherian, V., Browning, H., and Baritelle, J., History and cost of the pink bollworm (Lepidoptera: Gelechiidae) in the Imperial Valley, *Bull. Entomol. Soc. Am.*, 28, 286, 1982.

55. Staten, R. T., Flint, H. M., Weddle, R. C., Quintero, E., Zarte, R. E., Finnell, C. M., Hernandes, M., and Yamamoto, A., Pink bollworm (Lepidoptera: Gelechiidae): large scale field trials with a high rate gossyplure formulation, *J. Econ. Entomol.*, 80, 1267, 1987.

56. Staten, R. T., Miller, E., Grunnet, M., and Andress, E., The use of pheromones for pink bollworm management in Western cotton, in *Proc. Beltwide Cotton Prod. Res, Conf.*, New Orleans, LA, Brown, J. M., Ed., National Cotton Council, Memphis, TN, 1988, 206.

57. Bartlett, A. C., Genetics of the pink bollworm: sooty body and purple eye, *Ann. Entomol. Soc. Am.*, 72, 256, 1979.

58. Flint, H. M., Merkle, J. R., and Yamamoto, A., Pink bollworm (Lepidoptera: Gelechiidae): field testing a new polyethylene tube dispenser for gossyplure, *J. Econ. Entomol.*, 78, 1431, 1985.

59. Henneberry, T. J., Pink bollworm management in cotton in the southwestern United States, *USDA, ARS, ARS-51*, National Tech. Info. Service, Port Royal, VA, 1986.

Chapter 15

THE MOSCAMED PROGRAM: PRACTICAL ACHIEVEMENTS AND CONTRIBUTIONS TO SCIENCE

Dina Orozco, Walther R. Enkerlin, and Jesus Reyes

I. INTRODUCTION

The Mediterranean fruit fly, *Ceratitis capitata* (Wiedemann), a pest of great economical importance in the world, was first reported in the Western Hemisphere in 1904 in Brazil. Since then, it has spread throughout South and Central America. Only in a very few cases has its eradication been successful.

After its detection in Guatemala in 1976 and due to the economical importance that this pest represents for Mexico,[1] it was decided by the Mexican government to create a large scale program (MOSCAMED Program) in which the accumulated worldwide experience of the previous 20 years of medfly research was to be applied to prevent the pest from moving further north. For this, bilateral agreements were signed with both Guatemala (1976), and the U.S.A. (1977), and the Sterile Insect Technique (SIT) was chosen as the main control method.

The construction of the worlds largest mass rearing facility was started in 1978. In 1979 the first sterile flies were released, and by 1980 the facility achieved the goal of producing 500 million sterile flies per week.

The application of aerial bait-sprays to suppress wild populations, followed by the release of sterile flies, resulted in the gradual eradication of the pest from 1979 to 1982.[2,3,4] Since then, a sterile fly barrier was established at the Mexico-Guatemala border that has prevented the northward spread of the fly into Mexican territory.[5,6]

In this paper we describe what in our opinion have been the most important achievements in three aspects of the program: field operations, large scale mass rearing, and quality control.

II. FIELD OPERATIONS

The technology to detect and control medfly in Mexico was transferred mainly from the Joint FAO/IAEA Division and the U.S. Department of Agriculture. The basic technology was: (1) establishment of a trapping net to detect the presence of the fly, (2) the application of bait sprays, by air or ground, to control outbreaks or to suppress populations prior to the release of sterile flies, and (3) establishment of internal and external quarantines. In addition, the use of the SIT was proposed at a level that was not used before.

0-8493-4854-4/94/$0.00+$.50

209

Table 1
Eradication Stage (1979-1982)

| Zone | Detection Systems | | Control Systems | | Quarantine | | Public Relations |
	Trapping	Sampling	Bait-Spray	Sterile fly	Internal	External	
Suppression	Medium (0.50/km²)	Medium	Massive (Aerial)	None	X	X	+++
Eradication	Low (.25/km²)	High	Hot-spot (Ground)	High (2000-5000/Ha)	X	X	+++
Containment	Low (.25/km²)	High	None	Low (500-1000/Ha)	X	X	++
Post Eradication	High (1/km²)	Low	Hot-spot (Ground)	None		X	+
Free	Low +	None	None	None		X	

Post Eradication Stage (1982-1992)

| Zone | Detection Systems | | Control Systems | | Quarantine | | Public Relations |
	Trapping	Sampling	Bait-Spray	Sterile fly	Internal	External	
Containment	Low (.25/km²)	High	None	Low (500-1000/Ha)	X	X	++
Post Eradication	High (1/km²)	Low	Hot-spot (Ground)	None		X	+
Free	Low +	None	None			X	

The strategy was adapted and/or developed by the MOSCAMED Program to be utilized in a tropical ecosystem and under mountainous terrain with limited communications. From 1979 to 1982 the available techniques were combined according to the pest situation and the area under control was divided in four zones: eradication, containment, post-eradication, and free zone. This strategy was the key to gradually eliminate the medfly from the coast of Chiapas in Mexico.[2,5,6] Since then, a permanent barrier of 300,000 hectares in southern Mexico and southwestern Guatemala has been established to prevent the northward movement of the fly (Table 1, Figures 1 and 2).

During this time the basic technology was modified to optimize human and material resources. The most important techniques that have been developed and/or improved in field operation activities are:

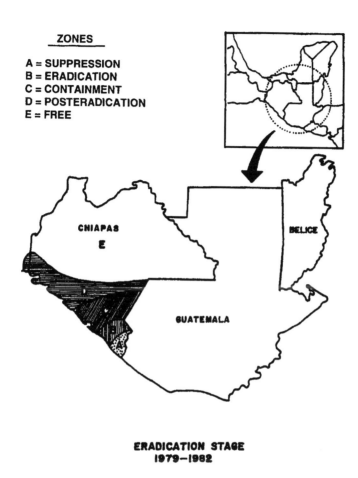

ZONES

A = SUPPRESSION
B = ERADICATION
C = CONTAINMENT
D = POSTERADICATION
E = FREE

CHIAPAS
E

BELICE

GUATEMALA

ERADICATION STAGE
1979—1982

Figure 1. Division of the working zones of the MOSCAMED Program in Mexico and Guatemala from 1979 to 1982. Eradication phase.

A.   AERIAL RELEASE OF STERILE FLIES

At the early stages of the program, there was not a developed technology to release such a large number of sterile flies. Several methods were proposed: "free pupae," "chilled adults," "paper bags," and "small boxes," with their respective advantages and disadvantages. Large field evaluations were carried out and because of the greater number of flies captured and low investment requirements, the "paper bag" method was selected.[7]

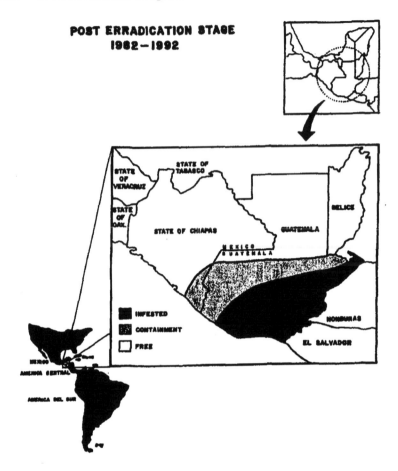

Figure 2. Current division of the working zones of the MOSCAMED Program in Mexico and Guatemala (1992). Containment Phase.

B.   FRUIT SAMPLING

By collecting all possible Medfly hosts in the State of Chiapas,[8] it was possible to identify the three primary hosts of the pest: coffee (*Coffee* sp.), star apple *(Chrysophyllum cainito)*, and guava *(Psidium guajava)*.[9] Knowledge of the seasonal infestation and the distribution of the fruit fly populations allowed us to schedule fruit sampling activities, increasing the possibility of detecting wild individuals. It was also found that in the case of isolated outbreaks with low population levels, fruit gathering was more efficient when fruit was collected from the tree rather than from the ground.[10,11]

C. **IMPROVEMENT OF TRAP DESIGN**

Following the FAO/IAEA research protocol for trap standardization, it was found that when a yellow insert is used in Jackson traps, it provides an additional visual stimulus to the trap which increases capture to about 60%, including additional females.[12]

D. **USE OF "LURE BAGS" WITH HYDROLYZED PROTEIN, MALATHION, AND TRIMEDLURE IN URBAN ZONES AND ECOLOGICAL RESERVES TO SUPPRESS WILD POPULATIONS**

The lure-bag consists of a bag made of rope (Mexican ixtle) (30x15cm) filled with tow and a hook. It contains a mixture of malathion, hydrolyzed protein, and water (1, 4 and 95%) and a trimedlure wick which is placed in the middle of the hook. This lure-bag operates as a spot of toxic bait which lasts in the field for about 30 days regardless of weather. This bag is selective because trimedlure repels other insects, including other tephritids.[13]

E. **REDUCTION IN CONCENTRATION OF MALATHION IN GROUND AND AERIAL BAIT SPRAY TREATMENTS**

Laboratory and field tests using different malathion concentrations in the bait resulted in reducing the amount of insecticide. In ground sprays a mixture of 57% malathion, hydrolyzed protein, and water, in a ratio of 1, 4, and 95%, respectively, is used. In the case of mango, for example, sprays are directed to foliage of the trees, applying four squirts of 50 cc each in alternated rows. The total amount of the mix applied in one hectare is about 4 liters. In aerial sprays the mixture is prepared with malathion ULV (96%) and hydrolyzed protein in a ratio of 10 and 90% respectively. This spray is applied over the canopy of the trees in alternated bands of 50 m wide. One liter of mixture is applied per hectare.[11,14]

F. **A "MICRO-FLUORESCENT" SYSTEM TO DIFFERENTIATE MARKED OF UNMARKED FLIES**

The methodology is normally used for the identification of sterile and fertile adult flies. The head is separated from the rest of the body of trapped adults and is crushed and examined in a chamber with ultra violet (UV) light. This causes the dye to fluoresce.[11] This method requires a large number of technicians with high error probability because of the many flies that have to be observed every day.

An alternative method to reduce labor costs and to increase efficiency is the use of a high resolution system, called "fluorescence system." It consists of a compound microscope that has a UV lamp integrated (Nikon Corporation 1990). By using the new system, separation and crushing of heads is not necessary in most cases because the complete fly body is observed through an external filter. If no fluorescent color is detected in the specimen, there is always the option of observing the specimen through the eyepiece (4X, 10X, 40X) which provides higher resolution. As an ultimate alternative, the process of crushing the head and observing it through the eyepiece can be used. The level of precision of this system reduces significantly the number of genitalia dissections that are done when flies have no mark.[15]

G. COLLABORATION WITH USDA-ARS IN THE DEVELOPMENT OF THE HOT WATER TREATMENT FOR MANGOES AS A POST-HARVEST CONTROL METHOD

The confirmatory test of the hot-water treatment for *Anastrepha serpentina* (Wied.) and the medfly was carried out using a water temperature of 45°C during 90 minutes. For both species, a population of at least 100,000 individuals was killed and a Probit 9 mortality level was reached.[16]

H. DEVELOPMENT OF A GAMMA IRRADIATION TREATMENT FOR MANGOES AS A QUARANTINE ALTERNATIVE (FOLLOWING GUIDELINES FOR FAO/IAEA AND ARS-APHIS-USDA PROTOCOLS)

A gamma radiation dosage able to induce a Probit 9 mortality level in *C. capitata*, *A. obliqua*, *A. ludens*, and *A. serpentina* was determined at the laboratory level in mangoes. The standard dose used in the confirmatory test was 10 krad for the *Anastrepha* species and 15 krad for *C. Capitata*. Although this did not result in death of the larvae, it was enough to impede adult emergence in a sample of at least 9,000 individuals for the first two species and 100,000 for the other two.

With the effective application of these technologies, among other things, we have been able to protect Mexico and the U.S. from the medfly for the past 15 years (Figure 3). This has prevented multimillion dollar economic losses due to direct damage to the fruit, as well as quarantine restrictions which would have stopped the marketing of fresh exported fruit.

III. LARGE SCALE MASS REARING

For the successful application of the SIT, the importance of producing large numbers of good quality sterile insects at a relatively low cost has long been recognized. The Metapa facility was the first large scale fruit fly mass rearing facility and still is the one with the highest weekly production.

Procedures at the Metapa mass rearing facility were initially derived from those developed at the FAO/IAEA Seibersdorf Laboratories in Austria.[17]

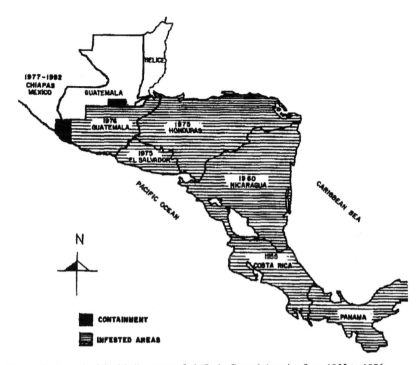

Figure 3. Spread of the Mediterranean fruit fly in Central America from 1955 to 1976.

Later, these procedures were modified according to local conditions and necessities to meet program requirements and reduce production costs. Modifications have continued through on-site research to achieve production goals and to solve unpredictable problems. The resulting system is currently known as the "Metapa System" and was described by Schwarz et al.[18]

Average weekly production at this facility during the past 12 years has been maintained over 500 million (Figure 4). The production costs per million pupae have ranged from $95.00 to $175.00 (U.S. dollars). The sterile flies have been distributed to Mexico, Guatemala, the U.S., and Chile. The distribution of the 1991 production is shown in Figure 5.

Here we briefly describe what we feel are the most important developments during the past 12 years:

A. REPLACEMENT OF THE WATER COLLECTING SYSTEM BY THE "DRY METHOD" FOR LARVAL SEPARATION

The diet is emptied into the larval separating machines to encourage the larvae to abandon the media. These machines, perforated with 1 cm

PUPAL PRODUCTION
1979-1992

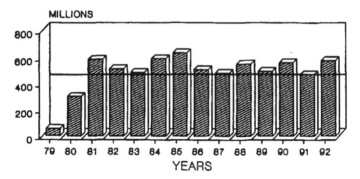

Figure 4. Average weekly pupal production at the MOSCAMED facility in Metapa, Mexico, from 1979 to 1992 (minimal requirement: 500 million/week).

PUPAE DISTRIBUTION
(MILLIONS)
1991

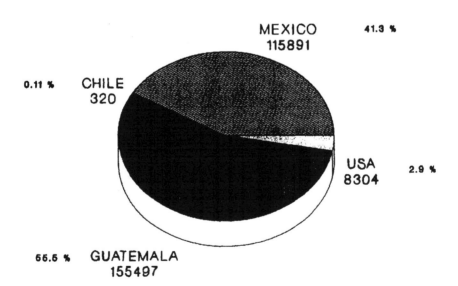

Figure 5. Distribution of the pupae produced at the MOSCAMED facility in Metapa, Mexico, during 1991.

diameter holes are lined with net cloth. This cloth retains the diet while allowing the larvae to escape. A collecting cloth funnel is located at the bottom of the machine into which larvae fall and are collected. Initially, larvae were collected in water but the "dry method" resulted in better recovery, and the quality of the flies was improved because of faster and more homogeneous pupation.

B.   NAKED PUPATION

The most common technique used for pupation is to add vermiculite or wheat bran to promote pupation. Instead of adding these materials, the mature pupae were placed on screened trays and transferred to a dark pupation room with 65-75% RH and 24-26°C. Normally 98-99% pupation is obtained in less than 24 hours. This modification avoided the need for separating the pupae from the pupation materials, and therefore, handling was reduced. This reduction in pupal handling resulted in better quality of the flies reared. It has been demonstrated that handling of mature larvae and young pupae has a strong adverse effect on the flight ability of the flies. This ability is very important for the flies to disperse and to find food and mating arenas.[19,20,21,22]

C.   DEVELOPMENT OF LARVAL DIETS

Locally available diet ingredients were tested to replace imported ones, and therefore, reduce diet cost and improve quality. Many diet formulas have been evaluated. After these evaluations two different diet formulas have been the most successful, they are known as the Sugar Beet Bagasse and the Texturized Soybean Flour diets.[22] The most successful larval diets used during these 12 years are shown in Table 2.

D.   DETERMINATION OF TEMPERATURE REQUIREMENTS

Temperature control in different rearing rooms was the most common problem at the facility. The study by Schultz[23] on metabolic heat production has led to a better understanding and management of the environmental conditions at the Metapa facility. The use of a data logger with terminal plugs in each room has permitted adequate control of temperature and humidity by means of two air conditioning systems.

E.   STARTER DIET AND RECYCLING DIET

The possibility of recycling the larval diet was investigated once the starter system was implemented for spent diets.[24] For this technique it was necessary to develop a starter diet in which the young larvae develop during the first two days. These techniques have been developed and tested under large scale conditions with satisfactory results.[25] If these techniques can be implemented, they will represent important reductions in rearing costs.

Table 2
Composition of Four Larval Diets for Mass Production of the
Mediterranean Fruit Fly at Metapa de Dominguez, Chiapas

INGREDIENT	BD$_5$	ST$_{50}$	SD$_{23}$	PD$_{25}$
		Percentage/Ton		
Yeast	9.96	8.50	6.70	8.60
Wheat bran	10.56	24.00	14.20	16.10
Sugar	8.97	8.20	7.30	8.30
Soybean		8.00	14.20	16.10
Nipagin	0.70	0.60	0.40	0.68
Benzoate	0.30	0.20	0.40	0.25
Beet bagasse	13.55			
Citric acid	0.70			0.70
Water	49.65	50.50	50.50	50.50
Total	100.00	100.00	100.00	100.00

IV. QUALITY CONTROL

Since the early stages of the Metapa facility, quality control was considered as an important aspect for sterile fly production. Again, from worldwide experience, a quality control manual was produced.[26] Recently, procedures to assess fly quality and set standards as agreed upon by the international community were incorporated to the facility protocols.[27] The annual pupal weight average, which is considered an important parameter to assess production quality, is show in Figure 6. Figure 7 shows the percentage of fliers (flight ability) obtained during these 14 years.

Quality control in insect mass rearing has been limited mostly to the evaluation on insect attributes. Less has been made on the control of the rearing process and the quality and efficacy of the reared insects under field conditions. The most desirable approach is to evaluate each step of the rearing process as well as the efficacy of the insects under natural conditions[28] During the past 5 years, in addition to the regular quality control tests, microbial, physical, and chemical specifications were established for raw materials[29] and a program on industrial relations was initiated. In this program the facility workers are encouraged to participate in the development and improvement of the rearing system.

During 1988, a total Quality Control Program was initiated including: data management (daily, weekly, monthly, and yearly reports), specifications

WEIGHT OF PUPAE
1979-1992

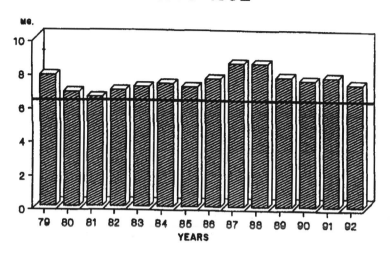

Figure 6. Average pupal weight of the sterile Mediterranean fruit flies produced at the MOSCAMED facility in Metapa, Mexico, from 1979 to 1992 (minimal value requirement 6.5 mg.).

PERCENTAGE OF FLIERS
1979-1992

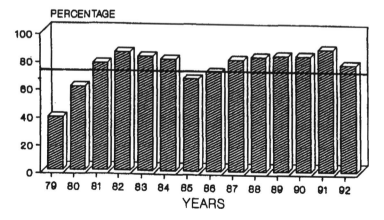

Figure 7. Flight ability of the sterile Mediterranean fruit flies produced at the MOSCAMED facility in Metapa, Mexico, from 1979 to 1992 (minimum requirement 75%).

of diet ingredients (physical, chemical, and microbiological aspects), quality control process (yields, efficiency, weight, etc.) and of course, the traditional insect quality tests: weight, adult emergence, flight ability, mating propensity, and longevity.

V. CONTRIBUTIONS

The most important contribution of the MOSCAMED Program to science was the eradication of the Mediterranean fruit fly from Mexico in 1982 and to keep the country free of this pest since then through the maintenance of an effective sterile fly barrier. This achievement demonstrated for the first time that a tephritid could be successfully controlled utilizing the SIT at a continental level in a region of high ecological diversity.

This technology and the accumulated experience has been transferred to more than 30 countries in Latin America, the Caribbean, Africa, and Asia by means of consultancy and training. A training course is annually sponsored at the MOSCAMED Program in Metapa.

Currently, institution efforts are dealing with the development of practical and accessible technology for developing countries. This technology includes the development of the SIT for *Anastrepha* species, the use of recycled diet, and use of parasitoids to suppress populations.

ACKNOWLEDGMENTS

We thank P. Liedo and C. O. Calkins for early reviews of the manuscript. We are grateful to J. Gutierrez Samperio for his continuous support and acknowledge support from the "Direccion General de Sanidad Vegetal, Secretaria de Agricultura y Recursos Hidraulicos."

REFERENCES

1.   **Gutierrez Samperio, J.,** La Mosca del Mediterraneo, *Ceratitis capitata* (Wiedeman) y los factores ecologicos que favorecerian su establecimiento y propagacion en Mexico, DGSV-SAG, Talleres Graficos de la Nacion, Mexico, D.F., 1976, 233p.

2.   **Hendrichs, J., Ortiz, G., Liedo, P., and Schwarz, A.,** Six years of successful medfly program in Mexico and Guatemala, in *Proceedings of the CEC/IOBC International Symposium on Fruit Flies of Economic Importance*, Athens, Greece, November 16-19, 1982, Cavalloro, R., Ed., A. A. Balkema, Rotterdam , 1982, 353.

3.   **Patton, P. T.,.** Mediterranean fruit fly eradication trail in Mexico, in *Proc. Symp. on Fruit Fly Problems. XIV Int. Congress of Entomology, Kyoto and Naha, Japan*, 1980, 18.

221

4. Patton, P., Programa contra la mosca del Mediterraneo en Mexico, in *Proceedings of a Symposium on Sterile Insect Technique and Radiation in Insect Control, Neuherberg, July 1981*, IAEA, STI/PUB/5952, 1981, 25.

5. **Ortiz, G., Liedo, P., Reyes, J., Schwarz, A. J., and Hendrichs, J.,** Mediterranean fruit fly (*Ceratitis capitata*): Present status of the Eradication Program in Southern Mexico. In: Cavalloro, R. (Ed.) Fruit flies of economic importance. 1984. A. A. Balkema, Rotterdam, 1986, 101.

6. **Schwarz, A. J., Liedo, J. P., Hendrichs, J. P.,** Current Program in Mexico, in *Fruit Flies: Their Biology, Natural Enemies and Control*. Volume 3B, Robinson, A. S. and Hooper, G., Eds., Elsevier, Amsterdam, 1989, 375.

7. **Villasenor Cortes A.,** Comparacion de tres sistemas de liberacion aerea para mosca del mediterraneo esteril, *Ceratitis capitata* (Wied). Tesis, Universidad Autonoma de Chiapas, Campus IV, Huehuetan, Chiapas, Mexico. 1985, 95p. (unpublished).

8. **Tejada, L. O.,.** Estudio sobre los hospederos potenciales de la mosca del Mediterraneo *Ceratitis capitata* (Wied) con enfasis en las presentes en el area del Soconusco, Chiapas, Mexico, Talleres Graficos de la Nacion, Mexico, 1980 95.

9. **Reyes, J. and Guillen, J.,** Hospereros de la Mosca del Mediterraneo *Ceratitis capitata* Wied. (Diptera: Tephritidae) en Mexico, *Folia Entomologica Mexicana*, 54, 19, 1982.

10. **Enkerlin, W. R. and Reyes, J.,** Evaluacion de un sistema de muestreo de frutos para la deteccion de *Ceratitis capitata* Wied., *Memorias Congreso Nacional de Manejo Integrado de Plagas (AGMIP).Guatemala, Centro, America,* 1984 (unpublished).

11. **Reyes, J., Villasenor, A., Ortiz, G., and Liedo, P.,** Manual de las Operaciones de Campo en una Campana de Erradicacion de la Mosca del Mediterraneo en Regiones Tropicales y Subtropicales, utilizando la Tecnica del Insecto Esteril. Programa MOSCAMED, DGSV-SARH. Tapachula, Chis., Mexico, 1986 (unpublished).

12. **Enkerlin W. R., Toledo, J., Reyes, J., Lopez, L., Villasenor, A., and Rios, E.,** Estandarizacion de Trampeo para Moscas del Mediterraneo en Programas que usan la TIE, *Memorias XXVI Congreso Nacional de Entomologia, Veracruz, Ver. Mexico,* 1991, 439.

13. **Enkerlin, W. R., Toledo, J. and Celedonio, H.,** Uso de bolsas-cebo para el control de la mosca del Mediterraneo *Ceratitis capitata* Wied. en zonas urbanas y reservas ecologicas, Programa MOSCAMED DGSV-SARH, Tapachula, Chiapas, Mexico, 1992 (unpublished).

14. **Enkerlin, D.,** Evaluacion de diferentes concentraciones de la mezcla de malation y proteina para el control de la mosca del Mediterraneo en la region del Soconusco, Chiapas, Mexico, Escual Practica. Instituto Tecnologico y de Estudios Superiores de Monterrey. (ITESM), 1985 (unpublished).

15. **Enkerlin, W. R., Lopez, L., and Celedonio, H.,** Evaluacion de un proceso alternativo para la diferenciacion de moscas esteriles de fertiles en *Ceratitis capitata* Wied. Programa MOSCAMED DGSV-SARH. Tapachula, Chiapas, Mexico, 1992 (unpublished).

16. **Sharp, J. L., Olye, M. T., Ingle, S. J., Hart, W. G., Enkerlin, W. R., Celedonio, H., Toledo, J., Stevens, L., Quintero, E., Reyes J., and Schwarz, A.,** Hot water quarantine treatment for mangoes from the state of Chiapas, Mexico, infested with Mediterranean Fruit Fly and *Anastrepha serpentina* (Wiedemann)(Diptera: Tephritidae), *J. Econ. Entomol.,* 82, 1663, 1989.

17. **Nadel, D. J.,** Current mass rearing techniques for the Mediterranean fruit fly, in *Proc. of a Panel on Sterile Male Technique for Control of Fruit Flies*, IAEA, Vienna, Austria, IAEA/STI/PUB/2, 1970, 13.

18. **Schwarz, A. J., Zambada, A., Orozco, D. H. S., Zavala, J. L. and Calkins, C. O.,** Mass production of the Mediterranean fruit fly at Metapa, Mexico, *Florida Entomol.,* 68, 467, 1985.

19. **Vargas, R. I., Chang, H. B. C., Komura, M. and Kawamoto, D. S.**, Evaluation of two pupation methods for mass production of Mediterranean fruit fly (Diptera:Tephritidae), *J. Econ. Entomol.* 79, 864, 1986.

20. **Orozco Davila D.**, Efecto del Manejo de las pupas como factor determinante de calidad en la cria masiva de Mosca del Mediterraneo, Tesis, Licenciatura en Biologia, Universidad Autonoma de Guadalajara. Guadalajara, Jal. Mexico, 1989, 62 p. (unpublished).

21. **Zavala, J. L.** Comparacion de procedimientos de cria de la mosca del Mediterraneo (*Ceratitis capitata* Wied.) en el Laboratorio de Produccion y Esterilizacion de Metapa de Dominguez, Chiapas, Mexico, Tesis, Ingeniero Agronomo Parasitologo, Universidad Autonoma de Chiapas, Campus IV, Huehuetan, Chiapas, Mexico, 1987, 97 p. (Unpublished).

22. **S.A.R.H.**, Programa MOSCAMED, Informes Anuales, Talleres Graficos de la Nacion, Mexico, 1980-84.

23. **Schultz, W. G.**, Thermodynamics of diptera-larvae rearing. A preliminary view, in *Fruit Flies of Economic Importance*, Cavalloro, R., Ed., A. A. Balkema, Rotterdam, 1983, 378.

24. **Bruzzone, N. D. and Schwarz, A. J.**, Recycling larval rearing medium in Mediterranean fruit fly mass production: A preliminary experiment, *J. Appl. Ent.*, 103, 418, 1987.

25. **Dominguez Gordillo, J. C.**, Uso de la tecnica de dieta larvaria iniciadora y reciclada como una alternativa en cria masiva de *C. capitata* (Wied) (Diptera Tephritidae), Tesis Licenciatura en Agronomia especialidad de Parasitologia, Universidad Autonoma de Chiapas, Campus IV, Huehuetan, Chiapas, Mexico, 1992, 79p.

26. **Orozco, D., Schwarz A., and Perez, A.**, Manual de procedimientos de control de calidad utilizado para evaluar la mosca producida en el laboratorio de produccion y esterilizacion de la Mosca del Mediterraneo, DGSV-SARH, Talleres Graficos de la Nacion, Mexico, 1983, 137p.

27. **Brazzel J. R., Calkins, C. O., Chambers, D. L., and Gates, D. B.**, Required quality control tests, quality specifications, and shipping procedures for laboratory produced Mediterranean fruit flies for sterile insect control programs, APHIS 81-51, Methods Development, PPQ, APHIS, USDA, 1986, 31p.

28. **Orozco D. and Lopez, R.**, Mating competitiveness of wild and laboratory mass-reared Medflies: Effect of male size, in *Fruit Flies: Biology and Management*, Aluja, M. and Liedo, P., Eds., Springer Verlag, New York, 185, 1993.

29. **Schwarz A., Orozco, D., and Liedo, P.**, Normas para la Adquisicion y Evaluacion de Ingredientes para Dieta Larvaria de Cria de la Mosca del Mediterraneo, 1989 (unpublished).

Chapter 16

THE MELON FLY ERADICATION PROGRAM IN JAPAN

Hiroyuki Kakinohana

I. INTRODUCTION

The melon fly, *Bactrocera cucurbitae* (Coquillett), not only causes serious damage to cucurbit crops but also infests various fruits and vegetables. Because of its occurrence in the Southwestern Islands of Japan, transportation of such hosts from the islands has been strictly prohibited by plant protection law. Such restriction makes it imperative to eradicate it from the islands.

In the Southwestern Islands there are two programs, Kagoshima and Okinawa Prefecture, aimed at eradicating the melon fly and lifting the ban on the transportation of host fruits. These programs are being carried out by the Prefectural Governments of Kagoshima and Okinawa, with the financial support of the Japanese National Government. The progress and the problems that occurred in these programs are reviewed.

II. DISTRIBUTION OF THE MELON FLY IN SOUTHWESTERN ISLANDS OF JAPAN

The melon fly was first discovered in Yaeyama Isls. in 1919. Ten years later it was found in Miyako Isls. In 1970, it was found on Kume Is., and spread to the Okinawa Islands in 1972. After that, the melon fly spread rapidly to the northern islands, i.e. Yoron and Okierabu islands in 1973, Tokunoshima, Amami-Oshima, and Kikai islands in 1974, and it was occasionally found on more northern islands (Figure 1).[1]

III. ERADICATION PROGRAM OF THE MELON FLY IN OKINAWA PREFECTURE

A. KUME ISLAND PROJECT

In 1972, an experimental eradication project of the melon fly using SIT began on Kume Island. This project was the first attempt to use this technique in Japan. Koyama[2] summarized the progress of the Kume Island project. The first step in the eradication project was suppression of the wild population with the male annihilation technique. From December 1972 to December 1974, suppressive control of the melon fly was carried out with

0-8493-4854-4/94/$0.00+$.50
© 1994 by CRC Press, Inc.

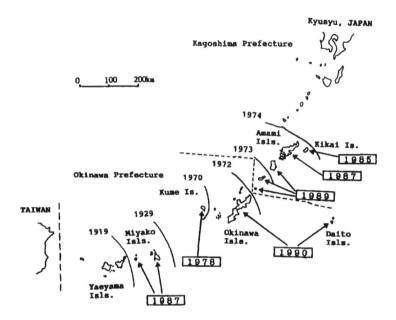

Figure 1. Distribution and eradication of the melon fly in the Southwestern Islands, Japan. Arabic numerals and that in squares show years of distribution and eradication, respectively. Broken line shows the border of the prefecture.

cotton rope or fiber block absorbed with cue-lure and insecticides, and with poisoned protein hydrolysate spray.[3] Since February 1975, about one million sterile flies were released as pupae in buckets weekly on Kume Is. However, the ratio of sterile and wild flies did not exceed one.[4] The estimation of the population density of the fly was carried out before the release,[5] and calculations based on an eradication model[6] indicated that 4 million flies were required to be released weekly to achieve eradication rapidly. From September 1975, the number of sterile flies released was increased to 2 million per week. As a result, the sterile:wild ratio became 10:1 in May 1976. The number of released flies increased to 3.5 to 4 million per week from May 1976, after which the sterile:wild ratio exceeded 100:1, and the percentage of infested host fruits decreased to zero in August 1976. Because the zero infestation level was maintained for one year, eradication had been considered to be achieved.[7]

B.   MIYAKO ISLANDS PROJECT

After successful eradication of the melon fly from Kume Is., a large scale project to eradicate the melon fly from Okinawa Prefecture was started.[8] In the first step, as a sub-project, the eradication project was started in the Miyako Isls. in 1984.

At first, Kuba et al.[9] attempted to estimate the density of male melon flies with the mark-recapture methods using the Hamada model, which is a modified Jackson positive method.[10] The peak number of male flies on Miyako Isls. was estimated to be 34.4 million in July. At that time, our mass-rearing facility could produce 30 million flies per week. Thus, Kuba et al. concluded that it was necessary to reduce the wild male population by male annihilation prior to the release of the sterile males.

For suppression with the male annihilation technique, cotton strings soaked in Cue-lure mixed with BRP (naled) and cut into 5 cm long pieces. These were distributed from helicopters at 40 strings per hectare once a month. At the final stage of the suppression, the wild male population was reduced to less than 5 percent of the average density before the control.

In August 1984, sterile fly releases were started on Miyako Islands. Thirty million flies were released each week evenly on the islands. In the initial 7 months of the SIT, the sterile fly population level was low (Figure 2). It was due not only to seasonal effects but also to a decrease in flight ability, i.e. droopy wing syndrome.[11] The fly quality was improved by sifting the pupae at a later stage after pupation. The distribution of the sterile and wild melon flies with Cue-lure traps on Miyako Isls. in August 1985 in Figure 3. Shimoji area is well known as a vegetable producing area in the islands. There were high density areas called hot spots[12] in some parts of Shimoji. Therefore, we decided to release an additional 6 million flies per week into the Shimoji area. The additional release area on Miyako Isls. is shown in Figure 4. The second additional release was carried out on the Ueno and Hirara area where new hot spots were detected. The third additional release was done to achieve the eradication as quickly as possible.

The results of the monitor trap survey (Figure 2) show that no wild fly had been caught since February 1987. Changes in the infestation ratio of host fruits on Miyako Islands is shown in Figure 5. Until April 1986, the range in the percentage of infestation was between 0.1 to 10. However, since May 1986, it rapidly decreased and it become zero after November 1986. From these data, it was concluded that the melon fly eradication on Miyako Isls. was achieved.

C.   OKINAWA ISLANDS PROJECT

Koyama et al.[13] estimated that the total number of mature male flies in Okinawa Island and adjacent islets at their most abundant season was about 38 million. However, the total number of wild flies reported by Koyama et al., was modified by the annual fluctuation and spatial distribution of population density caught by monitor traps. As a result, 110 million flies were estimated to occur at the peak density.[14] In Okinawa Islands (the largest island group in the prefecture), the eradication project was begun

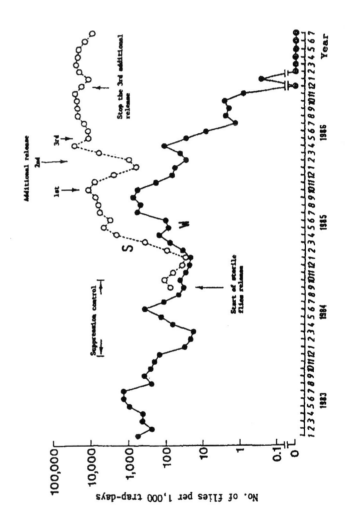

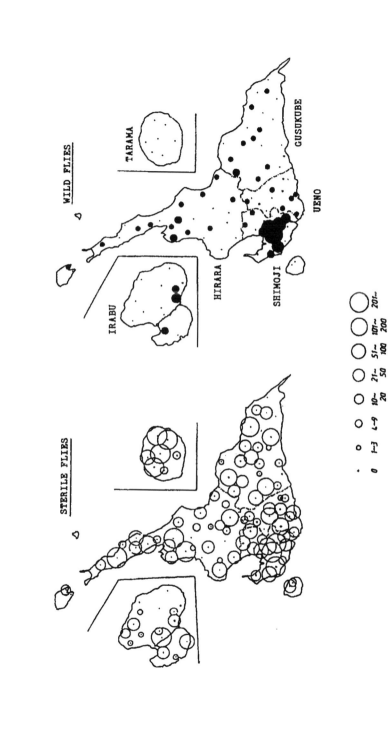

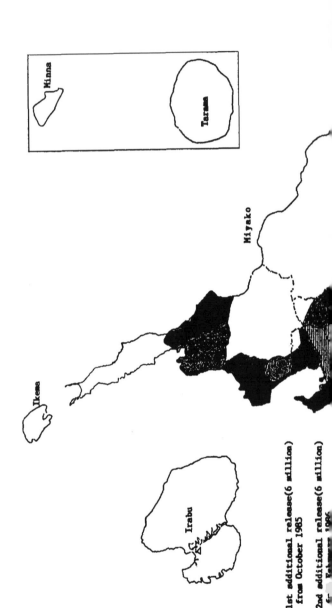

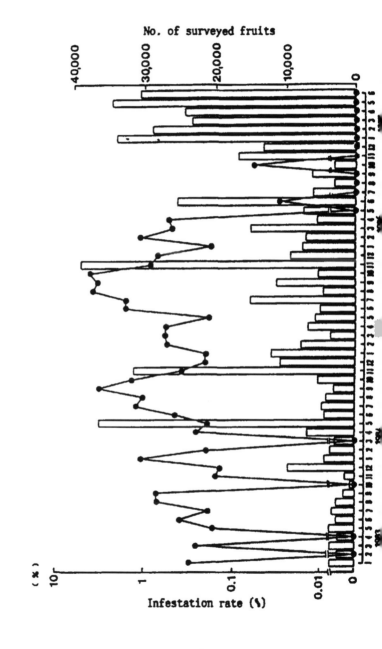

in 1986 while the program on Miyako Isls. was still in progress. Therefore, the control was initiated from the southern part of the islands, and the control area was expanded toward the northern part of the islands.

Before sterile fly release was started, suppression with the male annihilation technique was initiated in the southern part of Okinawa Island, where the highest population density occurred. In the final stage of suppression, the wild male population in the area decreased to about five percent of the control. Following suppression in these area, 87 million sterile flies were released weekly from November 1986. At the same time, suppression with male annihilation technique was started at the northern area of the islands except in the mountain area where population density was low. Male annihilation was continued until the start of SIT.

Progress in the SIT on Okinawa Islands is shown in Figure 6. At first, 87 million flies were released weekly in the southern part, from November 1986 (Figure 6, Area I). Secondly, 20 million sterile flies were released weekly in Area II from March 1987, except for the mountain area (Area IV). Thirdly, 40 million sterile flies were released weekly from January 1988 in Area III, except for the mountain area (Area IV). At the same time, 6 million sterile flies were released weekly on Daito Isls. Then, the release area was expanded to Area IV, but the number of released flies was the same (60 million) as the total number of flies released in Areas II and III. Finally, 7 million sterile flies were released weekly from June 1988 on adjacent islands in the northern area (Area V).

In the initial stage of SIT, sterile flies were released uniformly throughout the control area. Later, when our monitoring program detected the location of hot spots, we released additional flies around hot spots. These additional release were effective in achieving quick eradication of the melon fly. The results of the monitor trap survey on Okinawa Islands are shown in Figure 7. Survey data of traps and host fruits showed that no wild fly and no infested fruit were detected since December 1989 in Okinawa Isls. Based on these data, eradication of the melon fly on Okinawa Isls. was officially announced on November 1990.

D.   YAEYAMA ISLANDS PROJECT

The eradication project was begun in October 1989 in Yaeyama Islands, which is the final target area of eradication of the melon fly in the prefecture. After suppressing the males with the annihilation technique for about three month, 44 million flies were released weekly on Ishigaki Is. from January 1990. In the area of Yaeyama Islands, other than Ishigaki Is., the suppression was initiated in May; 43 million sterile flies were released weekly from November 1990. The survey data of traps and host fruits show that no wild fly and no infested fruit were detected since September and June

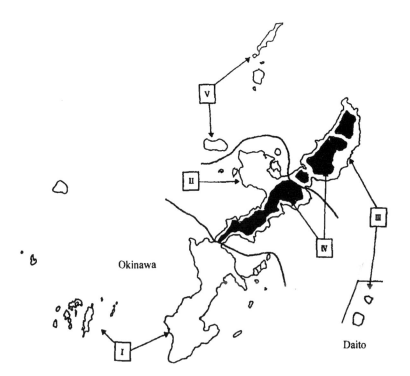

Figure 6. Progress of sterile flies release on the Okinawa Islands. I: 87 million flies per week were released from November 1986. II: 20 million flies per week were released from March 1987. III: 40 million and 6 million flies per week were released for Okinawa Is. and Daito Isls., respectively, from January 1988. IV: Release areas were expanded to all over the northern part of Okinawa Is. from II & III from April 1988. V: 7 million flies per week were released from July 1988.

1991, respectively. Eradication of the melon fly on Ishigaki Islands will be officially announced in 1993.

IV. ERADICATION PROGRAM OF THE MELON FLY ON KAGOSHIMA PREFECTURE

On Amami Islands in Kagoshima Prefecture, one of the Southwestern Islands, a decade eradication project of the melon fly using SIT began in 1979. Progress of the SIT on Amami Islands were reviewed by Fukushima and Tanaka,[15] Tanaka,[16] and Oshikawa et al.[17] The estimated population of wild flies on each islands is showed in Table 1.

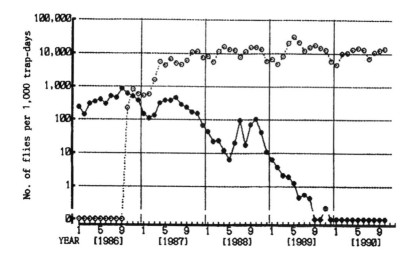

Figure 7. Monthly change in abundance of sterile and wild melon flies caught by monitor traps on the Okinawa Islands. Solid and open circles indicate wild and sterile flies respectively.

Before the sterile fly release, suppression of the wild population with the male annihilation technique was carried out by distributing fiber board squares and cotton ropes (both impregnated with lure-toxicant) or a bait spray of MEP-protein hydrolysate mixture. These treatments were applied one to three times a month until sterile flies were released.

The released number of sterile flies on Kikai Is., Amami-Oshima, Tokunoshima Is., Okierabu Is., and Yoron Is. were 4, 32, 13.6, 7.2-14.4 and 1.8-4.5 million per week, respectively. The result of the SIT on Amami Islands is shown in Figure 8. The eradication of the melon fly on Kikai Is. and Amami-Oshima were officially announced in 1985 and 1987, respectively. The melon fly was eradicated in 1989 also on Tokunoshima, Okierabu, and Yoron islands.

V. CONCLUSIONS

The schematic representation of the progress of SIT on Miyako Islands[18] is shown in Figure 9. In the first stage (I), suppression with the male annihilation technique was carried out. As cotton strings with a lure-toxicant mixture were distributed evenly, we expected the local population density to be reduced independent of the density. After suppression, in stage II, we started SIT. Sterile flies were released uniformly, but the density of sterile flies was lower than expected (broken lines). Thus,

Table 1
Estimated Wild Melon Flies Population on Amami Islands
(Modified from Fukushima and Tanaka)[15]

	Islands				
Items	Amami-Oshima	Kikai Is.	Tokunoshima	Okierabu	Yoron
Area (ha)	81,908	5,571	24,804	9,451	2,082
No. of Wild Flies (million)	50	4.6	11	5.4	1.2

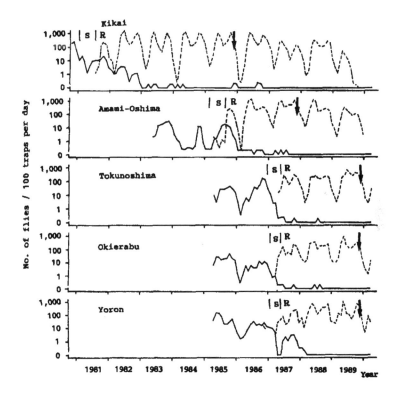

Figure 8. Monthly change in abundance of sterile and wild melon flies caught by monitor traps on the Amami Islands. Broken and solid lines show sterile and wild melon flies, respectively. S, R and arrows indicate the period of suppression, start of sterile flies release, and the time of eradication announced, respectively (modified from Fukushima and Tanaka).[15]

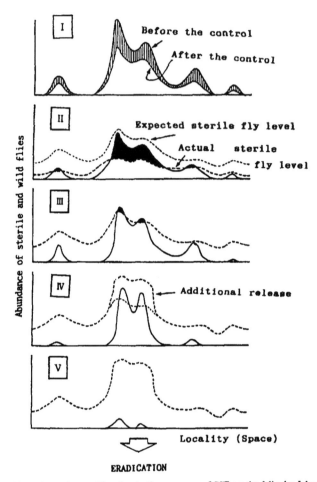

Figure 9. The schematic consideration to the progress of SIT on the Miyako Islands. I: Wild male population was reduced initially through the male annihilation technique. II: Sterile flies were released uniformly. However, sterile fly population level was not enough because of droopy-wing syndrome. High density areas of wild fly in certain parts become conspicuous. III: By qualitative improvement, sterile fly population was increased. However, it was insufficient to suppress the reproductive rate of the wild fly in hot spots. IV: By additional release, the ratio of sterile to normal (wild) of hot spots increased. V: The ratio of sterile to normal (wild) increased drastically, and eradication was achieved.

the numbers were insufficient to suppress the reproduction of wild flies in certain areas. The lower densities were caused not only by low quality (droopy wing syndrome) of released sterile flies but also by densities of wild flies that were high in a favorable reproductive environment, referred to as a hot spot. In this stage, sterile to normal ratios of all traps and a minimum sterile to normal ratio of a trap were less than 10 and 1, respectively. In

Stage III, sterile to normal ratios of all traps and minimum sterile to normal ratio of a trap were between 10 to 100 and 1 to 10, respectively. By qualitative improvement, the sterile fly population was increased. However, it was insufficient to suppress the reproductive rate of the wild fly in hot spots. Thus, we released additional flies into hot spots in Stage IV, and the ratio of sterile to normal (wild) of hot spots increased. In this stage, sterile to normal ratios of all traps and minimum sterile to normal ratio of a trap were between 100 to 1,000 and 10 to 100, respectively. In Stage V, sterile to normal ratios of all traps and a minimum ratio of a trap were more than 1,000 and 100, respectively. The ratio of sterile to normal (wild) increased dramatically, and we succeeded in the eradication efforts.

Our experience indicates that fly quality and additional releases are important. At first, we released sterile flies uniformly, however, we did not consider that the wild fly was distributed uniformly over the area. We expected that the distribution of sterile flies was going to create variable ratios to that of wild flies and thus will help us to identify the locations of hot spots. This occurred and we were able to release additional flies in the hot spots. The additional releases based on field trap data is good strategy for the success of SIT. Detailed field studies on the distribution pattern of wild flies are essential for success of SIT.

ACKNOWLEDGEMENT

The author wishes to express his thanks to Akira Tanaka, Kagoshima Agricultural Experiment Station, Japan, for permission to use their data in preparing this report.

REFERENCES

1. Ito, Y., Iwahashi, O., Kakinohana, H., and Sugimoto, A., The fruit fly problems in Japan (I), *Kagaku*, 46, 348, 1976.

2. Koyama, J., The Japan and Taiwan project on the control and/or eradication of fruit flies, in *Sterile Insect Technique and Radiation in Insect Control*, IAEA, Vienna, 1982, 39.

3. Iwahashi, O., Teruya, R., Teruya, T., and Ito, Y., Changes in abundance of the melon fly, *Dacus cucurbitae* Coquillett, before and after the suppression with cue-lure baits and protein-hydrolysate spray, *Jap. J. Appl. Ent. Zool.*, 19, 232, 1975.

4. Iwahashi, O., Ito, Y., and Zukeyama, H., A progress report on the sterile insect release of the melon fly, *Dacus cucurbitae* Coquillett (Diptera:Tephritidae). on Kume Island, Okinawa, *Appl. Ent. Zool.*, 11, 182, 1976.

5. Ito, Y., Murai, M., Teruya, T., Hamada, R., and Sugimoto, A., An estimation of population density of *Dacus cucurbitae* with mark-recapture methods, *Res. Popul. Ecol.*, 15, 213, 1974.

6. Ito, Y., A model of sterile insect release for eradication of the melon fly, *Dacus cucurbitae* Coquillett, *Appl. Ent. Zool.*, 12, 303, 1977.

7. Iwahashi, O., Eradication of the melon fly, *Dacus cucurbitae*, from Kume Is., Okinawa, with the sterile insect release method, *Res. Popul. Ecol.*, 19, 87, 1977.

8. Kakinohana, H., A plan to construct the new mass production facility for the melon fly, *Dacus cucurbitae* Coquillett, in Okinawa, Japan, in *Sterile Insect Technique and Radiation in Insect Control*, IAEA, Vienna, 1982, 477.

9. Kuba, H., Nakamori, H., Koyama, J., and Shinjo, G., An estimation of population density of the melon fly with mark-recapture methods, *Annual Report of Control Projects of Fruit Flies, Okinawa Prefecture*, Japan, 8, 188, 1983.

10. Hamada, R., Density estimation by the modified Jackson's method, *Appl. Ent. Zool.*, 11, 194, 1986.

11. Ozaki, E. T. and Kobayashi, R. M., Effects of pupal handling during laboratory rearing on adult eclosion and flight capability in three Tephritid species, *J. Econ. Entomol.*, 74, 520, 1981.

12. Shiga, M., Analysis of spatial distribution in fruit fly eradication, in *NATO ASI Series, Vol. G11, Pest Control: Operations and Systems Analysis in Fruit Fly Management*, M. Mangei *et al.*, Eds., Springer-Verlag, Berlin, Heidelberg, 1986, 387.

13. Koyama, J., Chigira, Y., Iwahashi, O., Kakinohana, H., Kuba, H., and Teruya, T., An estimation of the adult population of the melon fly, *Dacus cucurbitae* Coquillett (Diptera: Tephritidae), in Okinawa Island, Japan, *Appl. Ent. Zool.*, 17, 550, 1983.

14. Nakamori, H., Field survey and population estimation of the melon fly, *Dacus cucurbitae* Coquillett, in Okinawa, Japan, in *Proceedings of the International Symposium on the Biology and Control of Fruit Flies*, Kawasaki, K., Iwahashi, O., and Kaneshiro, K. Y., Eds., FFTC, Univ. of the Ryukyus and Okinawa Prif. Govern., 1991, Chap. I.A.II.

15. Fukushima, M. and Tanaka, A., Distribution and eradication of the melon fly on Amami Islands, in *Eradication of Fruit Flies - Theory and Practice*, Japan Agricultural Aviation Association, Tokyo, 1985, 317.

16. Tanaka, A., The progress of eradication of the oriental fly and the melon fly from the Amami Islands, *The Trace of Plant Protection*, A commemoration issue of the fifty year anniversary of the prediction of pest and disease density project and forty year anniversary of the plant protection project, Tokyo, 1991, 408.

17. Oshikawa, M., Enokizono, K., Torikai, M., and Kamifukumoto, A., Eradication of the melon fly, *Dacus cucurbitae* Coquillett, from the Amami Islands, Kagoshima, Japan, by the sterile insect release method, in *Proceedings of the International Symposium on the Biology and Control of Fruit Flies*, Kawasaki, K., Iwahashi, O., and Kaneshiro, K. Y., Eds., FFTC, Univ. of the Ryukyus and Okinawa Prif. Govern., 1991, Poster session, 338.

18. Kakinohana, H., Kuba, H., and Kawasaki, K., The eradication technique and problems of fruit flies, *The Trace of Plant Protection*, A commemoration issue of the fifty year anniversary of the prediction of pest and disease density project and forty year anniversary of the plant protection project, Tokyo, 1991, 466.

Chapter 17

THE ERADICATION OF THE QUEENSLAND FRUIT FLY, *BACTROCERA TRYONI*, FROM WESTERN AUSTRALIA

Kingsley Fisher

I. INTRODUCTION

The Queensland fruit fly (Qfly), *Bactrocera tryoni* (Froggatt), was detected for the first time in Western Australia during February 1989. Qfly maggots were found in tomatoes collected from a suburb in Perth. Until this discovery, the fly was kept out of Western Australia by interstate quarantine measures. In the eastern states of Australia, it is a major horticultural pest, attacking a wide range of fruits and vegetables.[1]

Western Australia has an incipient population of Mediterranean fruit fly (medfly), *Ceratitis capitata* (Wiedemann), that infests a large range of commercial fresh fruit.[2] However, the Qfly has the potential for becoming a more serious pest in Western Australia, being a more aggressive, more mobile, and longer lived pest than the Medfly.[3]

Existing control practices against the resident Mediterranean fruit fly are considered inadequate to cope with the additional pressure of Queensland fruit fly. The increased use of insecticides upon more crops would be expensive and socially sensitive. Markets overseas that accept fruit and vegetables from Western Australia on condition of its freedom from Queensland fruit fly status would also be threatened.

An eradication campaign, proposed in June 1989, began in September 1989 just as Queensland fruit flies were recovering from Perth's winter. Trap surveillance indicated that the infested area at the start of the campaign had expanded to more than 100 km² of suburban Perth.

A. ERADICATION METHODOLOGY

The campaign against the Queensland fruit fly consisted of a three-pronged attack: male annihilation, foliage baiting, and sterile insect release techniques.

Male Qfly annihilation began immediately after campaign approval. The method had shown significant suppression of this fruit fly on a number of other occasions.[4,5] The technique consisted of placing fibrous blocks (50x50mm), impregnated with the parapheromone Cue-lure (2 ml per block) and laced with maldison (2.3 g maldison/bl), on fruit trees and nonfruit trees. Male annihilation techniques were confined to the core of the infestation at the time, an area of 85 km², in an effort to contain the rate of spread of the fly. The majority of blocks were withdrawn before the commencement of releases of sterile flies.

Foliage baiting began within three weeks of the start of the campaign. A small work force of 300 people baited all fruit trees and, in some cases, non-fruiting shrubs within the infestation area. The foliage of each tree had approximately 50 ml of bait (0.9% of 1150g/L maldison, 5% protein autolysate) applied to it. The method had been used to suppress or eradicate fruit flies from commercial, urban, isolated, and semi-isolated communities for many years in Australia.[4]

The campaign of sterile Qfly releases was split into two sessions. The first (January-June 1990) commenced in January 1990 when sterile flies became available for release. This allowed the expensive chemical component of the campaign to be stopped and reduced the number of wild Qfly going into winter. The first session of releases stopped in June 1990 at the beginning of winter.

The second session (August-December 1990) commenced prior to when wild Qflies became sexually active in spring. At this time females needed to remate before they could lay fertile eggs.[6] During winter in Perth, the Qfly population was expected to be reduced to a lower level because of the cessation of mating and oviposition and the increased mortality due to cold. They may also experience restricted flight activity and the resorption of eggs and sperm (in spermathecae). These factors make it advantageous to release sterile Qfly towards the end of winter or prior to the onset of mating. The proliferation of sterility in the population is enhanced by the need of wild females to re-initiate their fertility and egg laying.

The Sterile Insect Technique has been widely used throughout the world for fruit flies and the success of large programs for medfly, melon fly, and oriental fruit fly eradication has won the technique considerable favor.[7,8] The production and release of sterile Qfly was based upon the successful 1984 eradication of medfly in Carnarvon, Western Australia,[9] a commercial fruit and vegetable growing region 1000 km north of Perth.

Within three months from the start of the campaign, a quarantine facility for mass rearing was designed and constructed. During this time a small colony of Qfly (200,000) was imported from research institutions in Queensland into a smaller quarantine facility in Perth. Based upon the Carnarvon experience, mass rearing, sterilization, and release techniques were refined for Qfly. Starting from a base colony of 2.5 million flies in December 1989, 35 million sterile Qflies/week were being produced by January 1990.

Eggs from caged flies were placed onto trays of a straw based artificial rearing diet and then placed into cabinets. The seeded medium was incubated in the closed cabinets at 26±1°C and 100% RH. The closed cabinets were removed to a cooler room (20±1°C) after 4-5 days to complete development when media temperatures reached 32-35° C. Cabinets were ventilated and the trays of media were watered (depending upon dryness) to assist in cooling and moisture control. Mature larvae popped from the medium during days 9-13 after egg-set and were caught in trays of water placed beneath the funnel bottom of each cabinet. Mature larvae were allowed to pupate using naked pupation.

Mature pupae were sterilized with 16 Krad radiation in 99.9% pure nitrogen atmosphere, 1-2 days prior to their emergence as adult flies. Quality control of sterile production was based upon the RAPID system.[10] Sterilized pupae were marked by mixing pupae with a fluorescent dye powder (5 g dye/litre pupae). Dyed pupae were measured volumetrically and placed into 45 litre plastic release bins. The interior of the bins had been roughened by sandblasting. Between 100 and 500 mls of dyed pupae were placed into each of 450-600 bins each day. A sheet of fluted newspaper was put over the pupae and several sugar cubes were added to each bin. Newspaper absorbed waste products and provided additional resting space for emerging flies. A ventilated wooden lid was made from marine 3-ply with a 100 mm hole in the center covered with fiberglass screen, designed to allow rapid release of the flies in the field.

The loaded bins were maintained at 27+1°C, 60+5% RH, and continuous light. As soon as the flies began emerging (36-48 hours), they were also provided with a 5% sugar solution in latex sponge blocks (75 mm x 75 mm x 35 mm) through the screened lid. The release bins were transferred to release sites in air-conditioned vehicles, 48 hours after flies began emerging.

An area received sterile flies at least every second day. The method of ground releases was simple and effective. It entailed taking a bin of flies to one of several hundred release sites, taking the lid off and allowing adults to fly away. Only flying adults were released. All unemerged pupae and empty pupae cases were returned to the laboratory and flies were allowed to emerge for an additional 24 hours. This ensured that all sterile released flies were thoroughly marked.

II. MONITORING AND ASSESSMENT

Monitoring and assessment of the campaign was based upon trapping techniques and upon fruit surveillance. Trapping was done using Lynfield traps[11] on an approximately 1 km grid (network) over 1400 km² of metropolitan Perth and in several hundred country towns in the southwest portion of Western Australia. The traps were serviced weekly. Before the discovery of Queensland fruit fly in Perth, no systematic trapping of the species was under way in Western Australia. Monitoring of Qfly using traps began in the Perth suburb of Dalkeith shortly after the initial discovery in February 1989 to delineate the infested area.

Some Nakagawa[12] traps were used in the areas with male annihilation blocks to supplement Lynfield traps in these areas. The Nakagawa traps were checked twice weekly to avoid putrefaction. Nakagawa traps were discontinued when sterile Qflies were released.

A supplementary trap system was used around any infestation, for the duration of the infestation. It was comprised of 5 Lynfield traps and 5 Nakagawa traps within 300 meters of an infestation and an additional 7 Lynfield traps were located within 500 meters. All supplementary traps were serviced twice weekly.

Marked sterile flies in traps were separated from unmarked flies using a microscope with a UV light source. Examination of trapped flies consisted of identifying dye particles in the sutures and on the cuticle and on the ptilinum in the head. The method described for releasing the sterile Qfly reduced the possibility of misidentifying sterile unmarked flies to less than 1 in 1000.

Fruit surveillance was done throughout the infestation area, until wild fly numbers were low in June 1990. Electrophoretic and morphological identifications were conducted to separate medfly and Qfly larvae and pupae retrieved from collected fruit samples.[13] One million individuals were examined during the course of the campaign. Fruit surveillance was particularly important in blocked areas and made several discoveries of larval infestations before trapping detected adults in these areas.

III. RESULTS AND DISCUSSION

The Queensland fruit fly was eradicated from Western Australia in December 1990. Eradication was confirmed after 12 months of intensive trapping during 1991. The campaign took 16 months to achieve eradication. It cost AUS$8 million and was the result of a concerted effort to apply as much pressure to the wild population as possible, using two chemical control techniques followed by the Sterile Insect Technique. The Qfly has not been found in Western Australia since its eradication.

Metropolitan Perth has 700 houses/km^2 with fruit trees in almost every backyard, highlighting the potential for the Qfly to become a serious problem. When the bait treatment started, the area being treated was slightly more than 100 km^2, but the area quickly grew to 300 km^2 during the spring. By December 1989, some 60,000 houses were receiving between 100-300 ml bait on a weekly basis (approximately 22,500 liters of bait per week).

While chemical techniques were suppressing the fly population during the spring when numbers were expected to escalate, the area of infestation

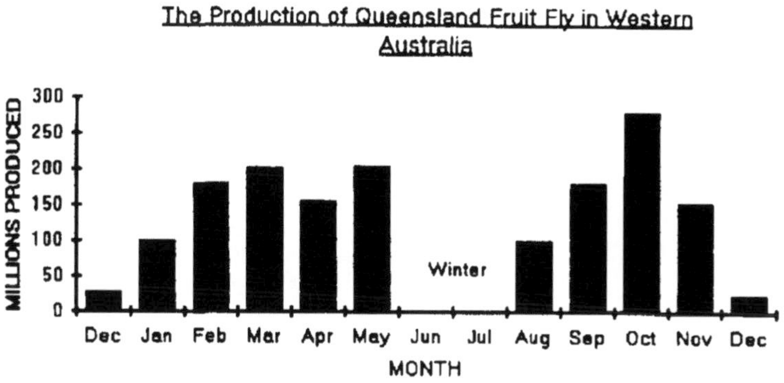

Figure 1. The monthly production of mass reared Queensland fruit fly during an eradication of Queensland fruit fly campaign in Western Australia (December 1989-December 1990).

TABLE 1
The Average Value of the Primary Components of the Quality Profile for Queensland Fruit Fly Mass Reared During an Eradication of Queensland Fruit Fly Campaign in Western Australia

Quality Test	Average	SE	n
Emergence	61.0%	2.20	241
Sterility (from residual egg hatch data)	99.8%	0.08	244
Mating Competitiveness (from 3:1:1 ratio test)	58.6%	0.02	31
Pupal Diameter (73% of pupae in 1.95-2.20mm size class)	1.90mm	0.50	140
Startle Activity Index	20.87	1.69	6
Longevity (time to 50% mortality of males only - with food and water provided)	45.7 days	1.46	35

was increasing. This effect of bait treatment, decreasing population density while leaving the population widely dispersed, was also noted in the eradication of the Medfly program in Carnarvon.[14]

Sterile Qfly releases began in January 1990. During 10 months, 1600 million flies were mass reared (Figure 1). A brief summary of the quality profile of the sterile Queensland fruit fly is shown in Table 1. High sterility and mating competitiveness characterized the sterile Queensland fruit fly. An emergence of 61% was unsatisfactory, but there was little time to adjust the

TABLE 2

The Number of Wild Queensland Fruit Fly

Caught in All Queensland Fruit Fly Traps in the Perth Metropolitan Area,

Before and During the Eradication Campaign

Month	Total # Traps in Release Zone (supplementary traps in brackets)	# Wild Qfly Trapped	# Sterile Qfly Trapped	# Wild Qfly/trap	# Sterile Qfly for Each Wild	Phase of Program
Apr 89	38(0)	158	0	4.158	N/A	Initial detection
May 89	253 (0)	100	0	0.426	N/A	and preliminary
Jun 89	235 (0)	26	0	0.111	N/A	trapping
Jul 89	0 (0)	0	0	N/A	N/A	
Aug 89	238 (0)	44	0	0.185	N/A	
Sep 89a	238 (0)	56	0	0.235	N/A	
Sep 89b	560 (231)	28	0	0.050	N/A	Start of eradication
Oct 89	717 (313)	20	0	0.028	N/A	campaign (Sep 5, 1989)
Nov 89	981 (449)	80	0	0.082	N/A	lure blocking
Dec 89	1197 (380)	88	0	0.074	N/A	and foliage baiting
Jan 90	1612 (433)	116	3688	0.072	31.79	Start of sterile releases
Feb 90	1576 (242)	41	41053	0.026	1001.29	and phasing out of

Jul 90	1269 (11)	0	790	0.000	N/A	
Aug 90	1269 (11)	0	58	0.000	N/A	
Sep 90	1269(11)	1	2566	0.001	2566.0	Spring session
Oct 90	1269 (11)	2	60847	0.002	30423.50	sterile releases
Nov 90	1275 (11)	6	158823	0.005	26470.50	
Dec 90	1230 (0)	0	92833	0.000	N/A	
Jan 91	1230 (0)	0	29063	0.000	N/A	
Feb 91	1230 (0)	0	373	0.000	N/A	
Mar 91	1230 (0)	0	13	0.000	N/A	
Apr 91	1225 (0)	0	0	0.000	N/A	
May 91	1225 (0)	0	0	0.000	N/A	
Jun 91	1225 (0)	0	0	0.000	N/A	Monitoring phase
Jul 91	1223 (0)	0	0	0.000	N/A	only
Aug 91	1292 (0)	0	0	0.000	N/A	
Sep 91	1292 (0)	0	0	0.000	N/A	
Oct 91	1292 (0)	0	0	0.000	N/A	
Nov 91	1292 (0)	0	0	0.000	N/A	
Dec 91	1352 (0)	0	0	0.000	N/A	
Totals		820	493980			

Table by courtesy of Mr. P. Yeoh

Notes: 1) N/A = not applicable

facility design to solve the problems. Of the 1.5 billion prepared for release, 950 million sterile Queensland fruit flies were estimated to be actually released.

The eradication of the Medfly project in Carnarvon[14] demonstrated the importance of a high frequency of low density releases to avoid sudden overpopulation that can lead to rapid dispersal of wild flies. It is important not to precipitate dispersal before mating has occurred. This release strategy also encourages a population of sterile flies of a wide range of maturities that are able to respond to the gradual emergence and presence of wild flies.

The frequent releases of sterile Qflies (almost daily) from January 1990 clearly suppressed the wild fly population in Perth during the first few months of releases. Wild Qflies had virtually disappeared going into the winter of 1990, as shown on Table 2, and that the second session of releases resulted in eradication by December 1990. Had a second session of releases not been done however, there was evidence that an outbreak of Qfly in August 1990 may have resurrected the wild Qfly population in Perth. The need for high overall ratios of sterile fruit flies to wild fruit flies for rapid suppression and eradication is illustrated in Table 2. For medfly, ratios of at least 1000:1 are considered necessary over a broad area to achieve economical eradication.[14]

For future protection of the Western Australian horticultural industry, the Department of Agriculture, with industry support, is continuing its surveillance for Queensland fruit fly using a 0.4 km trap grid containing over 3,500 Lynfield traps in the Metropolitan area and 350 traps in country towns. A strategic surveillance system has been adopted with risk profiling of habitats.[15]

IV. SUMMARY

An infestation of Queensland Fruit Fly was discovered in Perth, Western Australia, for the first time in February 1989. A program of sterile insect releases, foliage baiting, and male annihilation techniques was implemented in September 1989. Foliage baiting and male lure blocks were used to reduce the intensity of the infestation while mass rearing facilities were built. The area being treated in this way was 300 km². Both techniques ceased as sterile Queensland fruit fly were introduced into an area. The release of 40 million sterile Queensland fruit fly per week began in January 1990. All areas received sterile flies every alternate day and sterile releases covered an area of 800 km². The eradication of the Queensland fruit fly population was achieved within 16 months and has not been detected to date.

ACKNOWLEDGEMENTS

The success of the eradication of Queensland fruit fly campaign was the result of the effort of many people during 1989-1990, too numerous to acknowledge here. This report is on behalf of the Queensland fruit fly campaign team. However, to name a few, the work of the principal administrators; Dr. N. Monzu, Mr. A. W. Hogstrom, Dr. B. A. Stynes, and Mr. J. M. Bradshaw and of the other principal researchers; Mr. A. N. Sproul, Dr. D. K. Yeates, and Mr. P. Yeoh, is gratefully acknowledged.

REFERENCES

1. **Drew, R. A. I.**, Behavioural strategies fruit flies of the genus Dacus (Diptera:Tephritidae) significant in mating and host-plant relationships, *Bull. Entomol.Res.*, 77, 1987.

2. **Jenkins, C. F. H. and Shedley, D. G.**, The Mediterranean Fruit fly, *J. Agric. W. Aust.*, 2, 341, 1956.

3. **Drew, R. A. I.**, Fruit fly collecting, in *Economic Fruit Flies of the South Pacific Region.* 2nd ed., Drew, R. A. I., Hooper, G. H. S., and Bateman, M. A., Eds., Queensland Department of Primary Industries, Brisbane, 1982, 129.

4. **Bateman, M. A.**, Chemical methods for suppression and eradication of fruit fly populations, in *Economic Fruit Flies of the South Pacific Region.* 2nd ed., Drew, R. A. I., Hooper, G. H. S., and Bateman, M. A., Queensland Department of Primary Industries, Brisbane, 1982, 115.

5. **Cunningham, R. T.**, Population detection, in *Fruit Flies: Their Biology, Natural Enemies and Control*, Vol. 3B, Robinson, A.S. and Hooper, G.H.S., Eds., Elsevier, Amsterdam, 1989, 169.

6. **Meats, A.**, Strategies for maximizing the potential of the sterile insect release method: experiments with *Dacus tryoni*, *CEC/IOBC Symposium, Athens*, Nov. 1982, 371.

7. **Hooper, G. H. S.**, SIRM for suppression/eradication, in *Economic Fruit Flies of the South Pacific Region*, 2nd ed., Drew, R. A. I., Hooper, G. H. S., and Bateman, M. A., Queensland Department of Primary Industries, Brisbane, 1982, 98.

8. **Gilmore, J. E.**, Sterile Insect Technique. Overview, in *Fruit Flies: Their Biology, Natural Enemies and Control*, Vol. 3B, by Robinson, A. S. and Hooper, G. H. S., Eds., Elsevier, Amsterdam, 1989, 353.

9. **Fisher, K. T., Hill, A. R., and Sproul, A. N.**, Eradication of *Ceratitis capitata* (Wiedemann) (Diptera: Tephritidae) in Carnarvon, Western Australia, *J. Aust.Entomol.Soc.*, 24, 207, 1985.

10. **Boller, E. F., Katsoyannos B. I., Remund, J., and Chambers, D. L.**, Measuring, monitoring and improving the quality of mass reared Mediterranean fruit fly, *Ceratitis capitata* Weid: the RAPID quality control system for early warning, *Zeitschrift Fur Anglewandte Entomologie*, 92, 67, 1981.

11. **Cowley, J. M., Page, F. D., Nimmo, P. R., and Cowley, D. R.**, Comparison of the effectiveness of two traps for the Queensland fruit fly, *Dacus tryoni* (Froggatt) (Diptera: Tephritidae) and implications for quarantine surveillance systems, *J. Aust Entomol. Soc.*, 29, 171, 1990.

12. **Nakagawa, S., Suda, D., Urago, T., and Harris, E. J.,** Gallon plastic tub: a substitute for the McPhail trap, *J. Econ. Entomol.*, 68, 405, 1975.
13. **Dadour, L. R., Yeates, D. K., and Postle, A. R.,** Two rapid diagnostic techniques for distinguishing Mediterranean fruit fly from Queensland fruit fly, *J. Econ. Entomol.*, 85, 208, 1992.
14. **Fisher, K. T.** The successful eradication of *Ceratitis capitata* (W), the Mediterranean fruit fly, in Carnarvon, Western Australia, by the sterile insect method, in preparation.
15. **DeLima, F. and Yeoh, P. B.,** Future strategy for Qfly monitoring and eradication in Western Australia, in *Queensland Fruit Fly Eradication Campaign*, compiled by Sproul, A. N., Broughton, S., and Monzu, N., Western Australian Department of Agriculture Report, June 1992, 191.

INDEX

Printed and bound by CPI Group (UK) Ltd, Croydon, CR0 4YY

22/10/2024

01777605-0007